XRF Analysis of
Ceramics, Minerals
and Allied Materials

XRF Analysis of Ceramics, Minerals and Allied Materials

HARRY BENNETT
British Ceramic Research Association (Retired)

GRAHAM J. OLIVER
Ceram Research, Stoke-on-Trent, UK

JOHN WILEY & SONS
Chichester · New York · Brisbane · Toronto · Singapore

Copyright ©1992 by John Wiley & Sons Ltd,
Baffins Lane, Chichester,
West Sussex PO19 1UD, England

All rights reserved.

No part of this book may be reproduced by any means,
or transmitted, or translated into a machine language
without the written permission of the publisher.

Other Wiley Editorial Offices

John Wiley & Sons, Inc., 605 Third Avenue,
New York, NY 10158-0012, USA

Jacaranda Wiley Ltd, G.P.O. Box 859, Brisbane,
Queensland 4001, Australia

John Wiley & Sons (Canada) Ltd, 22 Worcester Road,
Rexdale, Ontario M9W 1L1, Canada

John Wiley & Sons (SEA) Pte Ltd, 37 Jalan Pemimpin #05-04.
Block B, Union Industrial Building, Singapore 2057

Library of Congress Cataloging-in-Publication Data:
Bennett, Henry, 1919-
   XRF analysis of ceramics, minerals and allied materials / Henry
Bennett, Graham J. Oliver.
      p.   cm.
   Includes bibliographical references (p.) and index.
   ISBN 0 471 93457 7 (cloth)
   1. Ceramic materials—Analysis. 2. X-ray spectroscopy.
I. Oliver, Graham J. II. Title.
TP810.5.B45   1992   92-8092
620.1'404295—dc20   CIP

British Library Cataloguing in Publication Data
A catalogue record for this book is available from the British Library
ISBN 0 471 93457 7

Typeset by Dobbie Typesetting Limited, Tavistock, Devon
Printed and bound in Great Britain by Biddles Ltd, Guildford, Surrey

Contents

Foreword xi
Preface xiii
   Acknowledgements xv

1 Introduction 1
  1.1 General 1
  1.2 Development of XRF Procedures 2
  1.3 Effect of XRF on the Reported Results 5
  1.4 Purpose of the Book 7
  1.5 Relation of XRF to Other Instrumental Techniques 8
  1.6 Contents of the Book 9
      References 11

2 Apparatus and Equipment 13
  2.1 Introduction 13
  2.2 Sampling 13
  2.3 Sample Preparation 13
      Jaw crushing, fine grinding, sample splitting
  2.4 Drying 21
  2.5 Weighing 22
  2.6 Heating 24
      Gas burners, furnaces
  2.7 Platinum Ware 27
      Size and shape, care and maintenance
  2.8 Selection of Spectrometer 31
      Type
  2.9 Computer 35
      Hardware, software
  2.10 Reagents and Blank Determinations 36

3 Determination of Non-XRF Elements 39
  3.1 Introduction 39
  3.2 Methods for Non-XRF elements 41
    3.2.1 Determination of Borate 41
    3.2.2 Determination of Sulphur 42
    3.2.3 Determination of Carbon 43

	3.2.4	Determination of Lithia	44
	3.2.5	Determination of Halogens	44
	3.2.6	Determination of Selenium	44
	3.2.7	Determination of Volatile or Alloying Elements	45
		References	46
4	**Loss on Ignition**		47
	4.1	Introduction	47
	4.2	Chemical Changes due to Heating	50
	4.3	The Effect of Composition	51
	4.4	The Evolution of Volatiles	55
		Water, carbon dioxide, sulphur gases, halogens	
	4.5	Conclusion	64
5	**Decomposition of Samples by Fusion**		67
	5.1	Introduction	67
	5.2	Choice of Flux	69
		Theory of suitability	
	5.3	Dilution of Sample in Flux	72
	5.4	Standards and Samples	75
		Standards	
	5.5	Casting	79
		Effect of weight of cast bead	
	5.6	Fusion of Non-oxide Materials	81
	5.7	Chrome-bearing Materials	85
	5.8	Problem Elements	88
		Reducible, volatile, concentration limited	
	5.9	Dilution	91
		References	93
6	**Selection of Instrument Parameters**		95
	6.1	Introduction	95
	6.2	X-ray Tubes	96
	6.3	X-ray Tube Parameters	97
	6.4	Instrument Masks	101
	6.5	Sample Masks	101
	6.6	Primary Collimator	102
	6.7	Crystals	102
	6.8	Detectors	103
	6.9	Flow Counter Windows	103
	6.10	Pulse-height Settings	104
		Reference	104
		Recommended Reading	104

CONTENTS vii

7 Element Line Selection ... 105
7.1 Introduction ... 105
7.2 The Eight Common Oxides ... 107
Silica, titania, alumina, ferric oxide, lime, magnesia, potash, soda
7.3 Other Oxides/Elements ... 113
Fluorine, phosphorus pentoxide, sulphur trioxide, chlorine; oxides of scandium, vanadium, chromium, manganese, cobalt, nickel, copper, zinc, gallium, germanium and arsenic; bromine; oxides of rubidium, strontium, yttrium, zirconium, niobium, molybdenum, cadmium, tin and antimony; iodine; oxides of caesium, barium, lanthanum, cerium, praseodymium, neodymium, samarium, gadolinium, ytterbium, hafnium, tantalum, tungsten, lead, bismuth, thorium and uranium

8 The Standard Procedure ... 145
8.1 Introduction ... 145
8.2 Determination of the Loss on Ignition ... 146
8.3 Preparation of the Bead ... 148
Criteria, fusion, casting, cooling, automatic preparation
8.4 Presentation of the Bead ... 153
8.5 The Remainder of the Method ... 153

9 Calibration ... 155
9.1 Introduction ... 155
9.2 Standards ... 157
9.3 Calibration ... 158
Establishing calibration curves, composition of standard beads, checking the calibration curves, presentation of standard beads, determination of line interference, determination of inter-element effects
9.4 Notes on Calibration ... 168
Determination of α-coefficients
9.5 Systematizing Calibrations and Fusions ... 169
Unifying calibration and fusion systems; base calibration: silica/alumina, limestone, high magnesia, zirconia, alternative approaches; standard methods

10 Presentation of the Sample Bead and Completion of the Analysis ... 177
10.1 Pre-presentation Treatment ... 177
10.2 Presentation of the Bead ... 178
10.3 Calculation of the Results ... 179
Correction for tungsten carbide contamination

11	**Routine Techniques for Material Types**...............	185
	11.1 Introduction..	185
	11.2 'Normal' Experimental Conditions	186
	Typical materials, sample preparation, drying, weighing, loss on ignition, fusion, casting, oxides to be determined, calibration ranges, reporting of results	

Chapters 12–17. The specific procedures for material types are described under the following headings:
   Typical materials, sampling, drying, weighing, loss on ignition, fusion, casting, constituents to be determined, calibration ranges, reporting of results.

12	**Procedures for Silica/Alumina Range Materials**	193
13	**Procedures for Calcium-rich Materials**	201
	13.1 Calcium Silicates	201
	13.2 Calcium with Silica and Phosphate (Bone China Bodies)	203
	13.3 Calcium/Phosphorus Pentoxide	204
	13.4 Calcium/Sulphur Trioxide...........................	206
	13.5 Calcium/Fluorine	208
	13.6 Calcium/Carbon Dioxide with/without Magnesium: Dolomite and Limestone	210
14	**Procedures for Magnesium-rich Materials**..................	215
	14.1 Magnesium Hydroxides, Oxides or Carbonates.........	215
	14.2 Magnesium Silicates	217
	14.3 Magnesia/Chrome-bearing Materials..................	219
15	**Procedures for Zircon-bearing Materials**....................	223
16	**Procedures for Various Oxides and Titanates**	227
17	**Procedures for Glasses, Glazes and Frits**	231
18	**Procedures for Reduced Materials**	239
	18.1 Introduction..	239
	18.2 Typical Materials....................................	240
	18.3 General Procedure	240
	18.4 Silicon-bearing Materials	242
	18.5 Silicon Carbide	243
	18.6 Silicon Nitride	245

18.7	Silicon (Metal)	246
18.8	Sialons	247
18.9	Ferro-alloys	247
18.10	Boron Carbide; Boron Nitride	248
	Reference	249

19 Procedures for Samples of Unknown Composition 251

19.1	Introduction	251
19.2	Typical Materials	252
19.3	Basic Procedure	253
19.4	Variants in Procedure	255
	Glasses and Glazes (see Chapter 17)	
	Vitreous Enamels, etc.	255
	Slags	258
	Kiln Deposits	259

Appendix I	**Loss on Ignition Techniques**	261
Appendix II	**Specific Fusion Techniques**	267
Appendix III	**Problem Elements or Oxides**	271
Appendix IV	**Certified Reference Materials**	273
Appendix V	**Laboratory Accreditation**	287
Index		291

Foreword

N. E. Sanderson

British Ceramic Research Ltd

Looking back at the foreword written by one of my predecessors, N. F. Astbury, in the last edition of a rather similar publication* concerned with mainly non-instrumental methods, I found that he began with the words:

> In a chemical world increasingly furnished with ingenious 'black boxes', contrived to take the human drudgery out of analysis, one is apt to overlook that these devices can function only if in support there is a soundly based practice of reliable direct chemical methods.

The above was written in 1971, a mere twenty years ago, and it is now clear that the sentiments expressed therein have been overtaken by events. Industrial analysis by 'wet' methods now only exists where the laboratory cannot afford or obtain any alternative. No research of note is being or has been carried out to improve the methods that are to be found in the above book. Almost all ceramic and similar analysis is, in fact, being carried out by those 'black boxes', viz. XRF, ICP, AAS and ion-chromatography. XRF even finds itself as a technique accepted for use as a standard method and is now almost universally relied on in the preparation of certified reference materials. Instrumental analysis thus now supplies its own basis.

Ceram Research has been involved in ceramic analysis since the 1920s and in XRF analysis since the early 1970s. Both the authors have been involved with XRF since the latter time and have between them over 60 years' experience of ceramic analysis. Such experience as they possess, together with the expertise from a broad-based commercial ceramic analysis laboratory, should be made available to help in standardizing techniques. This can only lead to better and less costly analyses.

It is pleasing to see Harry Bennett co-authoring another definitive work in the field of analytical sciences. He is now retired and the 'torch' has been passed to Graham Oliver.

I hope that this volume will, like its predecessors, prove to be a useful 'bench book' and will meet the needs of both routine and referee analysts. I wish it well.

---

*Chemical Methods of Silicate Analysis, H. Bennett and R. A. Reed, published for the British Ceramic Research Association by Academic Press, London and New York, 1971.

Preface

The virtual replacement of traditional chemical methods by instrumental techniques for the analysis of silicates and carbonates in ceramics and geochemistry left the 1971 version of the book on this topic* almost a historical relic. It is true that many of the methods therein remain as both UK and international standard methods, but they are retained solely for the use of smaller laboratories where instruments are not available.

X-ray fluorescence has had the greatest impact in this field, although both atomic absorption spectrophotometry and inductively coupled plasma spectrometry have their role to play. Ceram Research† has been in the forefront in the development of 'wet' chemical methods since the early 1950s, producing in collaboration with laboratories of the ceramic and steel industries the range of methods described in the above book. The organization, however, came late into the field of XRF analysis but nevertheless acted as the focal point in the UK Standard Method development. Ceram Research, both as a research organization and an accredited commercial testing laboratory, developed a considerable expertise in the use of XRF as a means of accurately analysing a wide range of non-metallic materials. The wide acceptance of the three editions on 'wet' methods, of which the above was the last, suggested that a similar book detailing expertise on the XRF technique might be equally useful.

This book is the result. It is essentially a distillation of the combined experience of the authors, who between them have been analysing ceramics for more than 60 years. It is not intended as an academic treatise on XRF; many of these already exist and cover the field far better than the present authors could presume to do. The data in this book is that culled by practising analysts in the course of their work, both at the bench itself and in organizing and running a commercial analysis laboratory handling 4000+ samples per annum ranging from simple clays to the most complex materials. Practicality is the keyword.

We have retained many of the old usages in nomenclature, both in terms of oxide names and in the few places where solution strengths are mentioned. Most technologists, certainly ceramic technologists, refer to the constituent oxides in

---

Chemical Methods of Silicate Analysis, H. Bennett and R. A. Reed, Academic Press, London and New York, 1971.

†Ceram Research, the trading name of British Ceramic Research Ltd, formerly the British Ceramic Research Association, is used for convenience throughout the book. Most of the development work described in the book was carried out while the organization was known under its original name.

a ceramic material by terms such as lime, magnesia, soda and potash. Also, oxides such as those of iron are referred to as ferrous and ferric. In addition many, if not most, practising analysts find the convenience of grams per litre or volume + volume for describing solution concentrations made from solid or liquid reagents respectively far more convenient in the laboratory environment than the more esoteric 'molar'. We believe this nomenclature is still the more practised and as this is, hopefully, a practical book we hope that this approach will find favour.

It must be stressed that this is not a book about XRF analysis techniques in general. It is about a single variant, analysis by the fused, cast bead technique such as has been accepted as the basis for British and international standard methods. There is no doubt that this method will stand comparison with any other approach and will produce results of referee standard whilst being used on a routine basis. The procedures described, once calibrated, will be capable of handling an extremely wide range of material types, most of them simply and rapidly. No serious discussion of energy dispersive systems is attempted as, in the authors' opinions, results are not of equivalent value.

The book is intended as a compilation of practical information on the analysis of mineral type materials and products by the fused, cast bead method. On the one hand there is a brief discursive approach concerning the necessary laboratory equipment and its use, the processes involved in loss on ignition and decomposition by fusion and their chemistry. On the other, most of the book is concerned in detail with regard to spectrometric parameters and the analytical procedures for a very wide range of materials. These, together with tables in appendices, allow easy reference to the parameters to be adopted to follow a defined procedure for most types of sample.

Many analytical books tend to approach individual determinations and analyses in a broadly discursive manner, often reviewing the strengths and weaknesses of each procedure in such depth and with so many reservations that the seeker for knowledge may well conclude that no such determination is possible. The approach here is different, being primarily to present methods in such a form that they will be accepted as descriptions of laboratory procedures by such organizations as NAMAS (the National Measurement Accreditation Scheme). As with previous volumes from the same laboratory, we believe that we have described a series of methods of proven value, as most have been in regular use for the best part of a decade in a busy commercial laboratory. Through our co-operation with the larger laboratories throughout the manufacturing and user industries, we know that essentially the same methods have been successfully used elsewhere. Therefore, we are confident that the methods can be applied to referee-standard analysis provided that the prescribed procedures are followed and adequate care is taken.

It is hoped that the compilation of information in this book will form a useful laboratory reference book. As such, it can fulfil a useful function in providing

PREFACE

the detailed written-down procedures required by quality assurance organizations such as NAMAS. The more general discursive material has been included in the hope that it may prevent others making some of the mistakes the authors have made. Even experienced analysts may find a crumb or two of information not previously encountered. We hope that in producing this book we may help towards standardizing procedures, equipment and software, which, in turn, should help to reduce costs and make for better inter-laboratory comparison of results. If this turns out to be the case we shall have been well rewarded for our efforts.

ACKNOWLEDGEMENTS

Although his efforts to turn the study and analysis of ceramics into a science date back many years, our thanks must go first to Dr J. W. Mellor for his inspiration. In similar vein, we also acknowledge the work of Dr A. T. Green for his creation of the first joint working committees to improve and standardize refractory analysis resulting from the needs of the Second World War, and their continuation afterwards into the period when it became possible to develop newer and better chemical methods. Both were Directors of Research of what is now Ceram Research. More recently, we must thank Dr D. F. W. James, retired Chief Executive of the above organization, for permission to use the information in this book and to publish it, also the present Chief Executive, Dr Neil Sanderson, for allowing Ceram Research to assist in its publication.

We gratefully acknowledge our debt to the British Standards Institution for permission to use the description of the method in a format very similar to that used in BS 1902.

Our thanks must also go to many of the staff of the analytical laboratories of Ceram Research itself, especially Mike Holmes, as well as to our collaborators on various Working Groups, who have contributed significantly to the work in establishing XRF, and the fused, cast bead technique in particular, as suitable for standardizing through BSI and thence to ISO, etc.

Finally, our thanks must go to Alan Ambrose for reading and making helpful comments on the manuscript during its gestation period.

1 Introduction

1.1 GENERAL

The analysis of ceramics, minerals, rocks, glasses, slags and related materials is not the easiest of analytical tasks. Prior to the development of modern and instrumental methods, such an analysis involved a long series of sequential separations and determinations. Of these one, two or even more components could be major contents demanding high levels of precision, skill and accuracy. Added to this was the need, in most cases, for decomposition by fusion, contributing considerably to the level of 'impurities' in the solution from which the separations and determinations had to be made. Until the advent of organic reagents, solvent extraction, flame photometry and spectrophotometry such an analysis was a daunting task, only successfully accomplished to a referee standard by a relatively few experts. Even modern 'wet' methods, although more readily giving good levels of accuracy after modest training, take too long for routine control purposes. Thus, it is not surprising that the realization of the possibilities of X-ray fluorescence was eagerly awaited.

Chemical methods of analysis of silicate, carbonate and similar inorganic materials, particularly those of use to the ceramic industries, formed the subject of a previous book by one of the present authors. This was *Chemical Methods of Silicate Analysis*, published by the then British Ceramic Research Association, and it has seen three editions. The first edition published in 1958 was written jointly by H. Bennett with a colleague, the late W. G. Hawley,[1] in response to numerous requests for various written procedures, which had then to be supplied in typescript. The second edition, written with the same co-author in 1965,[2] included procedures developed more recently than the 'classical' method. This was followed by a third edition with another colleague, R. A. Reed, in 1971[3] which incorporated a wide range of modern chemical methods.

These three editions gave detailed methods for a wide range of ceramic and geological materials. Most bulk-manufacture ceramic products and their raw materials originate from minerals in the silicate and carbonate rock classifications. These chemical methods still form the basis of current British Standard methods of analysis by 'wet' procedures for most refractory materials (BS 1902, Part 2). BSI has now published the first two Standard Methods for refractory materials[4,5] using XRF, dealing with high-silica and aluminosilicate materials by procedures developed by the then British Ceramic Research Association in cooperation with manufacturers' and users' laboratories, and which effectively form the basis of methods described in this book.

1.2 DEVELOPMENT OF XRF PROCEDURES

Among a great variety of possible reference sources, the published proceedings of the series of Ceramic Chemists' Conferences arranged by Ceram Research serve as a convenient general historical record of the developing interest in XRF as applied to ceramics and allied materials. These Conferences were intended to illustrate the current 'state of the art' of ceramic and allied analysis, and papers were given by leading ceramic analysts from manufacturers, users and the Research Association itself. From these it is possible to trace the development of the progressive modifications used to improve accuracy and precision. Early methods, not involving fusion, were little better than semi-quantitative, except in very specific circumstances where the physical and mineralogical state of samples was consistent. Later, when a fusion step was included, followed by grinding and briquetting, the situation, although improved, still did not fully solve the problem of adequate precision, except under ideal conditions.

It is significant that in seven conferences held during the years 1963 to 1984, on only one occasion was XRF not a topic. Even this omission is balanced by the fact that on two occasions two papers were given. As early as 1963, Hickson[6] from Pilkington Brothers Ltd reported on the use of XRF for the analysis of clays and fluxes, either by fine grinding of the material itself or after dilution by fusing in a lithium borate flux and then grinding the bead. In either case the powder was pressed into a pellet using an organic binder. The results quoted included some inter-element corrections. Two years later Orrell[7a] of Carborundum Co. Ltd reported results using a similar fusion technique followed by fine grinding of the melt and pressing into a pellet with a binder such as Movol. Lowe,[7b] from Pilkington Brothers Ltd, showed at the same meeting that no additional developments had taken place concerning ceramic materials; sand and silica bricks could be handled by grinding and pelleting, flat glass analysed directly or by simple polishing, and other materials were handled as described by Orrell.

There were no papers on the topic in the third Conference (1968), but two were given in the fourth in 1971. In the first of these, Ambrose[8a] of British Steel Corporation described, for the first time, the use of a fused, cast bead as a method of presenting the sample to the instrument. Cooperative work on this approach had been carried out by a BISRA (British Iron and Steel Research

INTRODUCTION

Association) Study Group wherein the chosen flux was lithium tetraborate at a dilution of 5 parts of flux to 1 of sample. Inter-element effects were handled either by correction procedures or by adding a heavy absorber (lanthanum oxide) to the flux. Quoted results were now approaching the quality necessary to make XRF a true competitor to chemical methods in terms of accuracy and precision for, at least, routine work. A parallel approach using the actual bead for presentation to the beam was described at the same meeting by Ashley[8b] of Pilkington Brothers Ltd whereby two fusions at dilutions of 5:1 and 10:1 were carried out (multi-dilution) and calculations were handled by an equation derived by Tertian[9] to obtain figures for the major constituents. Minor constituents were obtained by direct calibration using the more concentrated beads. A further significant step reported by both these authors was the use of synthetic standards.

At this point, the refractories industries under the auspices of Ceram Research set up a cooperative Working Group to study both these procedures, single fusion with the application of corrections and multi-dilution, in both cases presenting the fused, cast bead to the instrument. This Group included laboratories from the refractories, glass, cement and iron and steel industries together with raw materials suppliers. Inter-laboratory results[10] showed standard deviations of $\simeq 0.5\%$ for major contents, about 0.02–0.04% for titania, lime and potash and up to 0.1% for magnesia and ferric oxide. In the light of the then available analysing crystals, etc., the poor precision for magnesia is to be expected, as count rates were so low. The equally poor precision of the results for ferric oxide was surprising, as there seemed at the time to be no obvious reason for this moderately heavy element not to give results of the same quality as those for titania. This gave rise to the opinion amongst Working Group members that the iron was possibly not being dissolved in the melt like the other constituents, but rather dispersed in some form similar to a colloidal suspension. Later there appeared evidence that the most likely cause was slight reduction and subsequent loss of iron. Although the results for major consituents were not yet considered good enough for accurate work, there were strong indications that further improvements in technique and instrumentation should produce a method capable of virtually eliminating the use of 'wet' methods for ceramic analysis.

The year 1974 saw a further publication from Ceram Research XRF Working Group, this time dealing with the analysis of magnesites.[11] These materials were chosen because they were not only widely used refractory materials but also because they effectively required only the determination of impurities in a single-oxide matrix (i.e. magnesia), thus giving rise, it was hoped, to no significant inter-element effects. A successful method for the analysis of magnesites was hoped to be a first step towards the analysis of more complex materials. Results presented in the paper were competitive with, and in some cases even better than, those by chemical methods.

It is significant that up to this point all fusion and casting had been carried out in graphite crucibles, grinding off the bottom surface of the bead and polishing as necessary. The introduction of a 95% platinum/5% gold alloy for casting moulds and fusion dishes completed the transformation of XRF analysis from an interesting possibility to a fully practical analytical method. Results, in general, were as good as all but those of the very best chemical analysis, and in many cases better. That situation was admirably summed up by Ambrose[12] at the fifth Conference in 1974, at which he quoted acceptable results for diverse types of materials, high-silica, aluminosilicate and high-alumina materials as well as magnesites. He also showed single wide-ranging calibration curves for typical constituents such as alumina and ferric oxide in materials as varied as iron ores, iron ore sinters, slags, chrome-bearing refractories, magnesites, firebricks, cements, dolomites and silica bricks.

Although the above history applies only to work with which Ceram Research was associated, it nevertheless shows the general pattern of activity being followed by a wide variety of workers to establish the XRF method. Minor changes in technique remained to be made after 1974 but much of the improvement since has been due to instrumental developments, particularly in terms of analysing crystals, making the determination of light elements viable. After all, any method which would not yield acceptable results for soda, necessitating an additional decomposition for its separate determination, would lose much of its attraction. Additional types of X-ray tube target, e.g. rhodium and scandium, have also helped to increase sensitivity (and selectivity) where it was much needed.

Although a number of laboratories within the industry possessed XRF spectrometers, it was not until the quality of the results reached the levels of good routine analysis that Ceram Research purchased an instrument (Philips PW 1410). Then an investigation into the development of methods appropriate to high quality analysis of a wide range of ceramic and allied materials was carried out. Within a relatively short space of time, methods had been developed for almost all the common silicate, carbonate or oxide based refractory materials, starting with high-silica, aluminosilicate, aluminous and high-alumina materials. Methods followed for basic refractories such as magnesite, dolomite, limestone and chrome-bearing materials and then for the analysis of zircon, zircon-bearing refractories and zirconia (natural baddeleyite or products stabilized using calcium, magnesium or yttrium oxides).

Since then it has proved necessary to calibrate for additional elements in many of the types of material. Not only has nature proved more bountiful in its distribution of elements in the earth's crust than the classical chemical analyst found it convenient to admit, but the ceramic technologist has found merit in adding other oxides to common materials in order to enhance their properties. Now that analysis is so simple and inexpensive it is considered routine to monitor clays, etc., for such elements as phosphorus, zirconium, manganese, barium and sulphur.

INTRODUCTION

The scope of XRF has grown rapidly since the early days, when it was considered by most analysts to be a tool capable of routinely carrying out large numbers of repetitive analyses. It soon became clear that by using fused, cast beads the instrument may (in fact, ought) to be calibrated against standards made synthetically from pure oxides, carbonates, etc. This capability means that, among other things, it is a perfectly simple technique to use for the analysis of 'one-offs'. The process is, of course, considerably more expensive than referring to an already existing calibration. It demands the making of at least one, and more usually two or three, synthetic standards, but even so, when compared with the cost of carrying out a complex analysis by traditional methods (assuming this even to be possible), XRF analysis of such materials is remarkably inexpensive. In many of these situations, when no information is available concerning the composition, it is usually necessary to carry out a 'scan', i.e. a semi-quantitative analysis, to identify the elements present and to gain some idea of their relative amounts.

Acting along these general lines, methods have been developed for more complex materials, glasses, glazes, slags, vitreous enamels, welding fluxes, ferrites, calcium and other titanates, pottery and enamel colours amongst others. Many of these classes of materials started as one-offs but, as more of the type needed to be analysed, their analysis became translated to a routine basis.

As a result of the developments described above, the authors have established almost 50 calibration types and for one of these classes of material calibration covers over 50 elements. The principles of the methods described in the chapters of this book are applicable to almost any situation where inorganic natural products and artifacts of which they form the raw material need to be analysed. The basic method may even be applied to metals; admittedly they would have to be converted to the oxides first, not normally a very realistic approach, but the fact remains that it is possible and yields satisfactory results. The analysis of ferro-alloys, covered within the text, is analogous, and illustrates a practical product of this capability.

The process of standardizing these methods through BSI, CEN and ISO is currently being undertaken in cooperation with many of the laboratories of ceramic users and manufacturers, often operating as 'Working Groups' chaired by Ceram Research.

1.3 EFFECT OF XRF ON THE REPORTED RESULTS

The highly specific nature of the XRF technique as compared with chemical procedures has some very significant effects. The traditionally accepted compositions of materials in this field have been derived by the classical method or, more recently, by chemical methods such as those in BS 1902. Even Standard Methods, for commercial reasons, prescribe analytical procedures for only those oxides of interest and ignore the presence of lowish contents of other elements that can give rise to interferences. The latter are not normally determined nor their effects allowed for in the procedure, let alone taken account of in ceramic technology. Most chemical methods produce erroneous figures due to their lack of specificity whereas, given correct calibration and correction for interferences in accord with normal practice, XRF is highly specific.

The first effect to be noted is that, for the same range of determinations, analytical totals are lower when the analysis is carried out by XRF, sometimes markedly so. For the classical method an analytical total within the range 99.5–100.5% would usually be thought acceptable, for modern traditional 'wet' methods the range could well be 99.5–100.2%, but when XRF is the method of analysis the range might well be 99.2–100.2%. This difference reflects the fact that not all constituents are determined by any of the methods, but in the case of traditional methods some undetermined constituents will have increased the recorded figure(s) by interference effects, thereby increasing the analytical total.

An illustration of the sort of effect that a new technique can have on previously accepted figures may be made by citing the determination of alkalis in some early US Bureau of Standards samples. Presumably using the Lawrence Smith method,[13] two samples were recorded as having about 0.8% of Na_2O, but using flame photometry only about 0.08% of Na_2O was found. Investigations revealed that the discrepancy was almost certainly due to the presence of about 0.1% of Li_2O. The very low atomic weight of lithium vis à vis sodium meant that, whichever method of final determination was used, the original using chloride conversion to sulphate or the precipitation of sodium zinc uranyl acetate, the lithia content had a very significant effect on the apparent soda content.

In the classical method, the content of alumina was obtained as the residual figure of the ammonia group precipitate after deducting the contents of other co-precipitated oxides, usually only ferric and titanium oxides and possibly any co-precipitated silica being actually determined. The precipitate often also contained manganese oxide, phosphorus pentoxide, zirconia, etc., as well as any contamination from the reagents. In addition, less than precise precipitation conditions and the presence of a little carbonate in the ammonium hydroxide could add a part of the lime and magnesia. All these could be counted as alumina. Similarly with chemical BS type methods where alumina is determined by compleximetric titration, any element normally titrated under neutral conditions and not removed by the solvent extraction is included. This means that zinc, copper and a number of other heavy metals would be recorded according to their equivalent usage of EDTA. Again in the determination of lime and magnesia, also titrated with EDTA, this time under alkaline conditions, the presence of strontia and barium oxide, as well as any other of the elements titratable under the appropriate conditions, will be recorded, provided that they have avoided being complexed with triethanolamine and/or, in the case of lime, precipitation with the magnesium hydroxide.

If figures from these sorts of methods are compared with results by XRF, even with simple materials such as ball or china clays, then figures by XRF may be expected to show lower contents for at least alumina, lime and magnesia. More complex materials will offer more possibilities for this type of error. Even with feldspars the presence of the oxides of rubidium, caesium, strontium and barium is not unknown, and one or other may be present in significant amounts, so that XRF results will differ from traditional results. An additional complication concerning the determination of alkalis is that traditional decomposition normally depends on a hydrofluoric acid decomposition. Even with materials such as clays, usually assumed to be wholly decomposed, there may well be minerals that are not totally attacked. Certainly, in the case of high alumina content samples, the attack may effectively be more an extraction with sulphuric acid than a decomposition. In the relatively distant past, when the flame photometric method was being developed, this form of attack gave results comparable with those by the Lawrence Smith method, which had previously been accepted as correct, so that a strict evaluation of their ultimate accuracy was never deemed necessary. Thus, flame photometric results after HF decomposition have been universally accepted, but it is probable that the higher results by XRF are nearer the truth.

It will be clear from the above discussion that discrepancies will occur between 'wet' chemical methods and XRF but, more importantly, that XRF results should *inter alia* be the correct ones.

1.4 PURPOSE OF THE BOOK

The book is intended essentially as a practical tool for use by the analyst faced with the need to analyse samples of the above-mentioned type in quantity and having the use of an XRF spectrometer. As in the earlier works, the methods described are all familiar to the authors by personal experience and as such are known to yield reliable results. Methods as described in the book are already in widespread use in the UK industry and overseas, and will very probably eventually form the basis of a suite of British and international procedures based on the use of XRF.

As has already been indicated, in keeping with the philosophy of previous books to describe only methods within the authors' own ambit, references concerning research work are confined to activity with which they have been associated, either directly, with the co-operation of the ceramic and steel industries or through the series of conferences organized by Ceram Research. This is not to suggest that the great contributions made by other workers in the field are not appreciated, but rather that the limited needs of this book are amply covered by more localized references. The book, being intended to be entirely practical in approach, does NOT include descriptions of the principles of XRF, as these have already been more than adequately covered by a number of authors. If information of this sort is required, the names of two suitable books are given at the end of this chapter.[14,15]

It is assumed that many readers of this book will be concerned with analysing fully many of the materials covered, so that there will be a need for other complementary procedures. This demands that some brief information is given about possible techniques for the determination of elements and oxides not amenable to XRF determination.

The book not only gives practical procedures for the types of materials covered but also provides some background information about ancillary equipment and processes. Part of the book deals with this more general information that might be described as 'know-how', arising from the combined experience of the authors in ceramic and allied analysis. Thus, although the book may be described as essentially a practical laboratory handbook, it also contains some of the elements of a textbook.

1.5 RELATION OF XRF TO OTHER INSTRUMENTAL TECHNIQUES

The advantages of XRF in comparison with chemically based methods are so obvious that no time need be wasted on them. In all respects, accuracy, simplicity, speed and cost, XRF is so much better that comparison is pointless. Even when it is compared with other possible instrumental techniques it still has virtually all the advantages.

XRF starts with at least one major attribute; it is the only technique capable of giving precise results that does not demand the preparation of a solution. The manipulation from a fused melt to a total solution when dealing with silicate materials is problematical, slow and costly. The transition of silica in the sample from the alkaline matrix of the melt to the acidity of the solution is difficult to achieve without some polymerization, possibly as far as forming a gel and being effectively lost to the solution. As well as chemical methods, both AAS and ICP require the use of solutions, and samples for analysis thus have to pass through this process. This alone gives XRF a considerable edge in terms of simplicity, speed and cost as well as probable accuracy.

In addition to this, both AAS and ICP have basic precisions little better than 1% relative so that, for major contents, single sprayings cannot achieve results comparable with those obtained with a single count in XRF. For possible referee work, it is likely that duplicate solutions and internal standardization will be needed and, even then, multiple sprayings (at least two) of each solution will be necessary. XRF, with its ability to determine all significant contents in one

INTRODUCTION

run, can on most types of sample produce analytical totals approximating to 100%. This means that, for routine work on simple samples, only one bead is usually needed, the analytical total acting as a cross-check. If the need for regular replacement of standard solutions and the possible need to make up dilutions or separate solutions with additives to overcome interferences etc. for AAS and ICP analysis is added, their competitiveness is further reduced.

This is not to say that either of these techniques has no place in the laboratory. The small laboratory in this field, unable to justify the purchase of an XRF spectrometer, finds AAS very useful for routine analysis. ICP, and to some extent AAS, complements XRF in providing determinations for trace elements which could not be determined commercially viably by XRF. In addition, ICP is valuable for enabling determinations of boron and lithium to be made, the former being particularly useful for the analysis of glasses and glazes. The larger laboratory with high throughputs of a variety of work will need both XRF and ICP, but will undoubtedly use XRF wherever it is applicable.

1.6 CONTENTS OF THE BOOK

The book begins with a range of general information concerning the equipment and apparatus, and its use and care, that are necessary to complete a successful analysis by the XRF fused, cast bead technique. This will be familiar ground to the experienced analyst, but if it prevents an error or a disaster in a single laboratory its inclusion will have been justified.

XRF is, of course, not capable of determining all elements; possible approaches to most of these exceptions are outlined.

Prior to presenting a sample to the beam it is normally necessary to carry out an ignition of the sample, even if the loss on ignition is not to be determined. The sample also has to be decomposed by fusion to produce a uniform glass bead that can be presented to the instrument. Both processes appear to be relatively simple and, as the main problems with the use of the XRF technique are based on the physics of the excitation, etc., the chemistry involved in both the foregoing processes tends to be forgotten. Nevertheless, the chemical reactions involved in both can be very significant and can result in failure of the analysis. For these reasons, the problems and the appropriate chemistry are discussed at some length.

The next chapters are devoted to the spectrometer, firstly to the value of various tubes and their power settings, etc., detectors, crystals and other components. Line selection for the range of elements likely to be encountered in a very broad selection of these types of materials is also discussed in some depth.

The book then considers the actual experimental details of the method and takes the analyst through to the completed work. The British Standard Method for Aluminosilicates (BS 1902: Part 9: Section 9.1) is taken as the norm, but with some additional detail. In the same section, the whole process of calibration against synthetic standards is described. The formulation of the standards and the basis on which they are made are discussed and detailed. In the process of calibration it is not sufficient merely to prepare beads containing increasing amounts of the element being calibrated, one also has to take into account possible line overlaps caused by one element on the measured wavelength of another and the inter-element effects of one element on another caused either by enhancement or by absorption. All these interferences are, fortunately, subject to physical rules, and can at least in theory be calculated, but in practice commonly used ones, at least, are best derived experimentally from special standards or avoided altogether by using suitable short range calibrations. Finally, allowance has to be made for 'blanks', i.e. the presence of the elements to be determined in the flux, or possibly, more accurately, the variation of these amounts between batches of flux.

Included in the chapter on calibration is a discussion on how ranges of calibration may be extended, thereby reducing the number of individual types of sample and minimizing effort. A brief digression on the progress of national and international standardization will also be found in this chapter. Also integral with calibration is a description of the method for obtaining the final results from the raw experimental data.

The final section of the book gives full experimental details for specific types of material and provides a ready reference for analysing individual samples. The first chapter of this section establishes a format that, hopefully, will make for easy reference to the parameters needed for any specific material which are described in subsequent chapters. Appendices show data which has been discussed at length in the earlier text but is summarized there in tabular form to make for easy reference. Finally, there are included Tables of Certified Reference Materials (CRMs) indicating the range of these materials available from British Chemical Standards and the National Institute of Standards and Technology (USA). There are numerous other sources of CRMs, notably BAM (Bundesanstalt für Materialforschung und -prüfung) and IRSID (Institut de Recherches de le Sidérurgie Francais) in Europe and NIM (National Institute of Metals SA) in South Africa. These CRMs will not only fill the obvious function of being used as check samples to ensure that the spectrometer, or even the trainee analyst, is producing correct results, but will also indicate the potential composition of samples outside the normal experience of the analyst. The final appendix is a short appreciation of accreditation schemes, which may be of some value to laboratories contemplating joining an organization such as NAMAS.

REFERENCES

1. Bennett, H. and Hawley, W. G. (1958) *Chemical Methods of Silicate Analysis*, The British Ceramic Research Ltd, Stoke-on-Trent.
2. Bennett, H. and Hawley, W. G. (1965) *Chemical Methods of Silicate Analysis*, Academic Press, London and New York.
3. Bennett, H. and Reed, R. A. (1971) *Chemical Methods of Silicate Analysis*, Academic Press, London and New York.
4. *BS 1902: Part 9.1* (1987) British Standards Institution, London.
5. *BS 1902: Part 9.2* (1987) British Standards Institution, London.
6. Hickson, K. (1963) *First Ceramic Chemists' Conference*, Special Publication No. 43, p. 37, British Ceramic Research Ltd, Stoke-on-Trent.
7a. Orrell, E. W. (1965) *Second Ceramic Chemists' Conference*, Special Publication No. 50, p. 70, British Ceramic Research Ltd, Stoke-on-Trent.
7b. Lowe, R. S. (1965) *Second Ceramic Chemists' Conference*, Special Publication No. 50, p. 79, British Ceramic Research Ltd, Stoke-on-Trent
8a. Ambrose, A. D. (1971) *Fourth Ceramic Chemists' Conference*, Special Publication No. 72, p. 7, British Ceramic Research Ltd, Stoke-on-Trent.
8b. Ashley, D. G. (1971) *Fourth Ceramic Chemists' Conference*, Special Publication No. 72, p. 17, British Ceramic Research Ltd, Stoke-on-Trent.
9. Tertian, R. (1968) *Spectrochimica Acta*, **24B**, 447.
10. Oliver, G. J. (1972) *The Analysis of Aluminosilicates by X-Ray Fluorescence*, Report of the Analysis Committee: XRF Working Group, Special Publication No. 73, British Ceramic Research Ltd, Stoke-on-Trent.
11. Ambrose, A. D. (1974) *The Analysis of Magnesites by X-Ray Fluorescence*, Report of the Analysis Committee: XRF Working Group, Special Publication No. 88, British Ceramic Research Ltd, Stoke-on-Trent.
12. Ambrose, A. D. (1974) *Fifth Ceramic Chemists' Conference*, Special Publication No. 86, p. 25, British Ceramic Research Ltd, Stoke-on-Trent.
13. Smith, L. J. (1871) *Am. J. Sci. (3rd Series)*, 1, No. IV, 269.
14. Jenkins, R. and De Vries, J. L. (1965) *Practical X-ray Spectrometry*, Springer-Verlag, New York, 4th Impression.
15. Tertian, R. and Claisse, F. (1982) *Principles of Quantitative X-ray Fluorescence Analysis*, Heyden, New York.

2 Apparatus and Equipment

2.1 INTRODUCTION

If the laboratory is a member of NAMAS or similar organization, defined standards will be demanded of the performance of the equipment used and its maintenance. As there is a growing tendency towards international standardization of such schemes, it may be taken for granted that certain basic demands will be made. It is also certain that these demands will ultimately reach the highest level, as any country making a specific demand is bound to require its acceptance as a price for full international agreement, rather than have it claimed that it is prepared to lower its own standards in order to reach that agreement.

It may be taken for granted that any laboratory that seeks accreditation within one of these schemes (and the commercial value of membership is great) will need to ensure that its basic equipment conforms to the required standards. In addition, the equipment will require maintenance and confirmatory performance checking at approved intervals so as to maintain continuity of membership. Failure to complete the required checks within the time schedule specified may result in suspension of membership.

This chapter discusses some of the necessary equipment and its use, but briefly, as most analysts entering this field will be familiar with many of these aspects.

2.2 SAMPLING

The analyst usually receives the material after bulk sampling, i.e. in the form of a laboratory sample. The problems of sampling from bulk, whether from a quarry face, a stock-pile or a moving sample from a conveyor belt, are outside the scope of this book. Any involvement in this sort of activity should be preceded by a detailed study of specialist literature on the subject. Information relating to refractory materials and ores will be found in the appropriate BSI, CEN and ISO standards.

2.3 SAMPLE PREPARATION

Reduction of large samples, whether large in total weight or of large particle size, has to proceed in a number of steps. Progressive reduction in particle size must precede any reduction in the weight of sample taken for further processing. The particle size at any stage in the process must be such that the weight taken for onward processing is adequate to ensure that the portion obtained remains truly representative of the whole. The introduction of mechanical grinding has meant that an adequately large portion can be finely ground, hence ensuring a more representative sample.

It may be necessary to dry some samples before treatment; for example, clays received directly from the seam. Here a 40°C dryer is helpful as, for some applications such as the determination of soluble salts, this is as high a temperature as may be safely used. In fact, it may be necessary to dry some samples at no more than ambient temperature.

2.3.1 Jaw Crushing

Reduction in particle size usually starts in a jaw crusher (Fig. 2.1). Modern laboratory jaw crushers will reduce material to 3–5 mm diameter particle size. This is usually small enough for a representative portion of the sample to be taken sufficient to fill a 50 ml or 100 ml vial for a Tema or similar mill. The product from the mill is then normally fine enough to be coned and quartered to provide the final analytical sample.

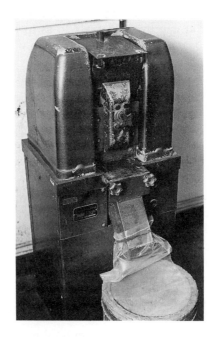

Fig. 2.1. Photograph of laboratory jaw crusher

APPARATUS AND EQUIPMENT

The jaw crusher handles fairly large weights of material and does not grind very finely. Contamination, even when grinding relatively hard materials, is not likely to be of great significance, except where particular analytical interest is centred on elements present in the metal of the jaws. This would apply in two types of case: first, where the sample contains a low content of iron and where its accurate determination is important, and second, where trace element determinations are required. Most types of jaws are constructed of highly alloyed steel and contain amounts of several elements other than iron. Experience seems to indicate that manganese steel jaws are amongst the most satisfactory for general use.

It is theoretically possible to remove much of the metal introduced during jaw crushing with a magnet, and this has frequently been recommended. Two factors militate against it. First, it is not possible to remove all the tramp metal, as some of it will be smeared on to individual grains unless the magnetic field is so great that it removes the grain itself. This second problem can also manifest itself if the sample itself contains magnetic material, a not unlikely occurrence.

Some cross-contamination is also very likely, as some of the fine material will be carried into the crevices behind the jaws, etc. The ease of cleaning between samples is a prime factor in the choice of jaw crusher when the laboratory is called on to analyse a variety of types of sample. Where it is merely desired to monitor the quality of a single product, cross-contamination may be of minor importance, but if several different types of material are being handled, contamination of a subsequent sample by traces of the previous sample can be disastrous. Minimizing this can be a very time consuming process, as the jaw crusher normally has to be partially dismantled so as to clean the machine thoroughly by brushing and vacuum cleaning. Planning the programme of grinding can make a significant contribution to the reduction of cleaning time by ensuring that, as far as possible, similar types of sample are ground sequentially. Where this does not apply it may be possible to reduce cross-contamination if, after normal cleaning of the machine, a portion of the sample to be treated is first crushed and then discarded. The crusher is then cleaned again and the remainder of the sample crushed.

2.3.2 Fine Grinding

The use of XRF has made possible the almost universal application of tungsten carbide as a grinding medium, as it is now easy to measure the level of contamination and make allowance for it. This very hard material has the advantage that the level of contamination is relatively low compared with most of the alternatives. Only boron carbide produces less contamination, but its use is severely restricted by its fragility. In practice, it can currently only be used in the form of mortar and pestle. It would be a great advantage if boron carbide were available on a device similar to a mechanical agate mortar and pestle. It is a material that merits more attention from the manufacturers of grinding equipment, as contamination is virtually ignorable in most instances. Tungsten carbide, on the other hand, although not so wear-resistant, is usually adequately so, and it is also tough, making it, if not ideal, at least very acceptable as a material for grinding vials. Tungsten occurs only rarely in ceramic or geological samples, so the material offers no problems in this respect.

Assuming that mechanical grinding is to be the norm, then it is very difficult to improve on a mill such as the Tema mill (Fig. 2.2), which has given 15–20 years service in the authors' laboratory without any but minor ailments. The machines are, by definition, noisy, so that it is very advisable to purchase at the outset a version with sound-proofing.

The only vials of interest to the mineral or ceramic analyst are those lined with tungsten carbide or agate (Fig. 2.3). The material used as lining for the former can vary quite considerably in physical characteristics. The manufacturer can usually provide material ranging from very hard, wear-resistant but tending

Fig. 2.2. Photograph of Tema mill

APPARATUS AND EQUIPMENT

Fig. 2.3. Photograph of tungsten carbide and agate vials, Tema mill

to be brittle, to a softer but tougher material. The best compromise for grinding vials is somewhere between the two extremes, and the choice of material should be best known to the experienced supplier. Additionally, the exact clearance between the vial lid and the puck and ring is critical; too small a clearance and grinding will be impaired, possibly even to the point of failure. On the other hand, too great a clearance means that chipping of the puck and ring, particularly the former, will occur much sooner in its life, possibly before there is any pronounced visible wear. Finally, the main body of the vial is of steel with only a tungsten carbide lining, and the lid comprises a tungsten carbide insert in steel. This makes it necessary to check the ease with which the material being ground can come into contact with the steel and, particularly, whether the material that has been in contact (and therefore contaminated) can return to the main body of the sample.

In practice, the nominal 50 ml capacity tungsten carbide vial seems to be of most general use. It is not desirable to grind too little sample in a large vial, as self-attrition of the tungsten carbide is more pronounced, greatly increasing the contamination. On the other hand, too much sample in too small a vial will be slow to grind even to the point of failure. Slow grinding will again enhance the probable levels of contamination. This contamination will not be solely tungsten carbide itself, as the lining material is usually bonded with cobalt, or occasionally with nickel, at about the 10% content level. Other impurities, in particular titanium and tantalum, are often present to an extent that may be relevant in trace determinations of the elements.

There is always a danger of some slight cross-contamination from one sample to the next, even after normal cleaning of the vial. In situations where this may be important, the technique of grinding a small portion of the sample and discarding it prior to the actual grinding greatly reduces the problem.

Most ordinary materials take only one to two minutes to grind to an analytically fine powder. This is not to say that the whole of the material is at a fineness of, say, less than 125 μm (BSS 120 mesh), but that almost all will be finer than that figure and the small amount of coarser material will still be fine enough to dissolve in the flux.

Harder types of material leave very little residue in the vial after brushing with a small, stiff brush. Softer materials such as clays and, especially, basic materials containing large amounts of lime or magnesia, such as magnesite, dolomite or limestone, tend to pack into the vial very solidly, making it very difficult to remove the whole of the sample. The difficulty can, in the case of clay at least, be reduced by ensuring that the material is thoroughly dried before attempting to grind it, but this technique appears to do little for basic materials.

These conditions make vial cleaning a problem. An easily removed material, after brushing, requires the vial only to be thoroughly washed with hot water, followed by a rinse with industrial or isopropyl alcohol to inhibit rusting, and then a wipe round with a clean cloth or tissue. If the sample has stuck, it is often necessary to add a little water and carry out a wet grinding for about 2–3 minutes, adding an appropriate amount of clean, washed sand (about 8–16 mesh). The residue is then rinsed away and, if necessary, a further wet grinding with sand is performed. When the vial appears clean, washing and drying can proceed as above. No evidence has been obtained to suggest that the material left in the vial is different in composition to that which comprises the sample.

When the vial is next to be used for an analysis of a sample in which one of the common oxides (e.g. silica) is an important 'trace', at least one pre-grinding with part of the sample is highly desirable. If there is insufficient sample to make this possible, grinding at least once using a material of similar type will usually suffice.

Even though tungsten carbide is both tough and hard, nevertheless wear becomes visible after a time. Eventually the lid and, to a lesser extent, the base of the vial will show worn rings. At the same time, the radius on the chamfer on the top and bottom of the puck and ring will have been worn away, tending towards a square corner. The slight additional play allowed by the wear on the top and bottom faces of the puck, ring and the vial itself allows impacts to occur at a slight angle, i.e. on to a corner rather than on face to face. This, together with the increasingly sharp corners, eventually causes the puck and/or possibly the upper edge of the lining to chip. This can produce a relatively massive contamination.

Once this happens, or when it is suspected that the level of wear is reaching this point, experience shows that, in most cases, simply replacing the puck and ring will enable the vial to be used for a further period of time. These replacements can last almost as long as the first set. However, when the same point is being reached again, a second simple replacement of the puck and ring will not usually suffice. At this point it is desirable to send the vial back to the suppliers for re-grinding. This involves skimming off the bottom surface of the carbide from the bowl and the inner side of the lid to produce smooth, flat surfaces. The ground out vial is then fitted with oversize puck and ring to match the slightly increased dimensions of the bowl. This puck and ring can again be replaced as before, but after this the vial is best discarded.

Agate vials have proved to be very disappointing. Although the Tema mill utilizes a slower rate of vibration for the agate vial, great care has to be taken in its use. Experience has shown that the rings supplied are very brittle, and it would appear that both the amount of sample and its type have to be taken into account. It is unfortunate that one mistake causes a broken ring, and as these are very costly it is inadvisable to experiment too often. It would seem advisable to avoid the agate Tema vial, and to use the much slower, but potentially less expensive, mechanical agate mortar and pestle.

Agate is only useful for grinding softer materials, and for samples where some silica contamination is not important or where one cannot afford to add any other elements. Even so, even agate is not pure silica; it has been known to contain inclusions of heavy metal minerals. This consideration normally would be significant only if small trace contents are sought in the sample used for general analysis.

Each type of sample, and the information required from its analysis, merits its own special consideration with respect to sampling. The consequences of the method selected for sample preparation have to be considered and steps taken to minimize the significance of contamination. In the most difficult situation it may even be necessary to prepare, separately by different routes, two entirely different portions of the sample, one for the main part of the analysis and a second to enable a particular element to be determined. The results of the main analysis then need to be re-calculated, to allow for the correct figure for the contaminant, by using the figure obtained from the second portion.

2.3.3 Sample Splitting

Large volumes are best handled by means of a sample splitter (Fig. 2.4), usually described as a 'riffle', after a preliminary jaw crushing. Where weights of 5 kg or less are to be split, it is probably better to turn to the use of a rotary splitter, such as the Pascall sample splitter (Fig. 2.5). This is essentially a conical hopper mounted over an upwardly mounted cone with six containers round the circumference of the base of the cone. The base holding the cone is rotated by means of an electric motor. This is intended to reduce the possibility of bias,

Fig. 2.4. Photograph of sample splitter (riffle)

Fig. 2.5. Photograph of Pascall sample splitter

as the material could tend to split in one direction. By rotating the containers each receives its share of each type of material.

2.4 DRYING

Pre-fired and most 'hard' materials, after grinding and transferring to a glass sample tube (75 × 25 mm or 50 × 25 mm), may be adequately dried at 110°C in 2 h. Most clays require 4 h. It is usual, unless there is reason to the contrary, to dry samples overnight at 110°C. A good quality drying oven (Fig. 2.6) is necessary, preferably with fan-assisted air circulation to ensure reasonably even heat distribution, and with true thermostatic control. As the oven is expected to be in service for many years, the additional cost of a stainless steel internal lining is justified. It may be necessary to have more than one drying oven, as some materials used for making standard beads, together with some types of sample, may require drying at other temperatures. The actual work-load of the laboratory determines the desirable sizes and numbers of ovens. It is better to err on the side of too large ovens than too small; work-loads have a habit of

Fig. 2.6. Photograph of drying oven

increasing. The temperature at various points within the oven should be checked at regular intervals.

Recent work with microwave ovens suggest that these may prove useful for drying many, but not all, types of ceramic samples.

Prolonged drying is better avoided, as in some instances it may be deleterious: two examples may be given. Some clays, particularly bentonitic clays (montmorillonite), will tend to lose water slowly but progressively at 110°C; this factor causes confusion between drying and ignition losses. Some other materials, such as quicklime, are liable to absorb water vapour and/or carbon dioxide from the atmosphere. Such materials may demand special treatment to ensure that this does not occur. One possible solution is to weight two portions of the sample at the same time, using the first portion for the analysis and the second for a determination of loss in weight on drying. Some materials need special drying conditions, for example, lower temperatures, and some in fact should not be dried at all but analysed as received; examples of the latter are gypsum and colemanite.

Standard beads are best stored in drying cabinets at 30–40°C to retard their deterioration by absorption of atmospheric water.

2.5 WEIGHING

After drying, the hot sample needs to be cooled in a good desiccator over a suitable drying agent. Usually silica gel is adequate, but those types of samples that tend to pick up water from the atmosphere strongly may need better desiccants. Commercial silica gel is very convenient; it is solid, a change of colour from blue to pink indicates its exhaustion, and it can be regenerated easily by simple drying at 110°C. Alternative and generally more effective desiccants include calcium chloride, phosphorus pentoxide and Anhydrone (magnesium perchlorate). The last needs to be used with care, as explosions have been reported in the literature from its use in the presence of organic materials.

Balances should be capable of a high level of accuracy; they need to be of good 'analytical' quality, nominally described as 'four-figure' balances (Fig. 2.7). 'Nominally' is used in this connection, as most practising analysts appreciate that if weighing is to be not worse than ± 0.1 mg it would be necessary to use a 'five-figure' balance. The generally used 'four-figure' analytical balance in good condition is capable of achieving tolerances of about ± 0.2 mg.

Few laboratories attempting XRF analysis would be equipped with any type of balance other than automatic weight-loading mechanical types or newer electronic balances. Mechanical balances with the advantage of automatic weight-loading and, possibly, constant tare made up to the late 1970s and described as four-figure analytical balances were ideal for the purpose. Unfortunately this no longer appears to be true. Then, most good four-figure

APPARATUS AND EQUIPMENT 23

Fig. 2.7. Photograph of four-figure analytical balance

single-pan balances loaded weights down to 10 mg, leaving only this amount to be divided on the optical scale between (usually) 100 divisions. Thus, one scale division is equal to only 0.1 mg and an error in sensitivity of 1% produce maximum errors of the same magnitude. More recently, balances may have as much as 1 g on the graticule, which would result in a much greater error if the sensitivity deteriorated.

Electronic balances appear to have several advantages over their mechanical equivalents. Sensitivity can be assessed rapidly, being done normally on at least a daily basis. Newer type electronic balances have a built-in calibration weight to enable simple and rapid recalibration; this is an important factor in terms of quality assurance. It is also very easy to set a tare; this can be a great advantage when weighing out sample or flux. The electronics of the balance allow the weights to be recorded and stored by a computer, enabling weighing errors to be readily detected and the weights stored for future reference. They can then be used for calculating losses on ignition and ultimately, by connection with the spectrometer system, in the analysis.

Electronic balances have their disadvantages, however. First, they are much more expensive, being anything up to twice the price of the equivalent mechanical balance. However, as they may have no realistic competitors in the future, this may have to be accepted. Another less obvious problem lies in the difficulties that can arise when adding increments of powder to the pan. With a mechanical balance it is usually a simple matter to assess the distance from the desired weight and progressively reduce the speed of addition of material to the pan as the correct weight nears. An electronic balance tends to produce over-swing after each addition by an amount that is almost random, not seeming to be connected with the weight added. This has been a point on which the manufacturers have focused, and more recent models incorporate damping systems which appear to be solving the problem.

2.6 HEATING

2.6.1 Gas Burners

The type of gas burner used will depend on local conditions and the type of domestic or bottled gas available. Generally speaking, Meker or Amal burners will be of most service. A device designed by Ceram Research (Fig. 2.8), consisting of a four-burner unit which is mounted on a rotating tilted platform driven by a small electric motor, provides a very convenient way of swirling the fusion dish and contents so that the fusion can be left to proceed without supervision. Many more sophisticated automated devices are available. The greater proportion of analyst time concerned with bead preparation is taken up at this stage, and if one relates the cost of this particular device to that of most other pieces of automation it must be very cost-effective. It is particularly

Fig. 2.8. Photograph of four-burner swirling device

applicable to the laboratory with a wide-ranging work-load, rather than to a situation which involves large numbers of essentially repetitive analyses. In this latter situation a much more expensive but more complete automation may be viable. This simple swirling device is available through Ceram Research*, to whom enquiries may be made.

The platinum ware used for loss on ignition determinations or fusions should not be allowed to come into contact with base metal while it is hot. Thus, the triangles used to support it above the burner should be of an inert, heat-resistant material. Silica-sleeved metal wire triangles are the most serviceable; with reasonable care these will last for a long time, failing eventually by flaking. They are easily damaged if they are allowed to come into contact with the flux, the melt or other alkaline materials such as sodium carbonate.

If a swirling device, such as the one mentioned above, is employed, it is necessary to make a grid out of a suitable type of wire and silica sleeves so as to fit the frame over the burners and to allow the dishes to sit correctly over the burners. Four-sided sleeved frames in the shape of a square are better than a more conventional triangular form. The frame is sited so that the base of the dish, when sitting firmly in the cradle, is just above the top of the blue cones of the gas flame. As the dishes can vary slightly in size and shape when they have been in use for a while, it is advisable to have a loop of wire, sheathed with silica where it will come into contact with the dish, under the dish so as to prevent its slipping through the holder.

There are gas burners (blast burners), utilizing either compressed air or oxygen, that will provide an adequate temperature for run of the mill samples, but at the cost of generating excessive heat and noise in the laboratory. Automatic fusion equipment based on gas can be very satisfactory for undemanding samples but, unless additional higher temperature equipment is available, restrictions will be imposed on the laboratory in terms of the range of samples that can be analysed.

2.6.2 Furnaces

Careful use of gas burners can, for a limited range of suitable samples, yield complete fusions and satisfactory beads. However, many samples demand higher decomposition temperatures than can be obtained with normal gas burners. In addition, if gas is used throughout, there is always the danger that oxidizing conditions may not be adequate to ensure total oxidation of lower oxides such as ferrous iron or Fe_3O_4. For these reasons, completion of the fusion in an electric furnace is preferable.

Electric furnaces are safer and much more convenient, as well as being more reliable for atmosphere and temperature control. Modern electrical equipment

---

*British Ceramic Research Ltd, Queens Road, Penkhull, Stoke-on-Trent, ST4 7LQ.

utilizing temperature controllers (not proportional controllers) provides an ideal environment for this work. By fitting furnaces with reliable fail-safe switches, so that when the door is opened the current is automatically turned off, they offer a degree of safety much greater than can be obtained with gas appliances.

It has been found most convenient and economical to use solely 1400°C furnaces with silicon carbide heating elements. The elements in these have a life of several months, used 24 hours per day, seven days per week, even at the higher temperature (1200°C) needed for fusion work. Generally, it is not the high temperature but the alkali vapours in the atmosphere that appear to cause deterioration. The use of the same type of furnace for all the work gives the laboratory a great measure of flexibility, in that each furnace can be transferred from one operation to another with minimum delay. This is particularly useful when a furnace burns out. It has the additional advantage that the type of replacement bar is common to all furnaces.

Heating elements, whether made from wire or from silicon carbide, once they have been heated to over 1000°C, tend to become brittle if they are allowed to cool below about 700°C. For this reason, it is advisable to maintain the furnaces at least at or above the latter temperature, even when they are not in use, e.g. at weekends or even during holiday periods. The cost of the electricity used, together with the additional wear on the elements, is more than compensated for by the increased life of the elements. Failure usually occurs in just one element, but experience has shown that it is better to replace all the elements. How far these element problems apply to light-weight kilns that can be brought to high temperatures in a matter of minutes remains to be ascertained but, as the materials used in the heating elements are similar to those of conventional furnaces, it is probable that the same problems apply. If this is so, the apparently great advantages of the light-weight kiln in saving electricity by turning it off overnight will be nullified. Recently developed microwave furnaces appear to have considerable potential in terms of flexibility and the capability of being turned off when not in use.

There are great advantages to be gained from using a tunnel kiln for the loss on ignition determination (Fig. 2.9). This enables the dishes, partly covered, to be placed just outside the entrance to the furnace. The slowly moving chain carries the dish and contents into the heating zone, where the temperature gradually increases to 1025°C over a period of 20 min. The hot zone of the kiln is of such a length to maintain the dish at this temperature for 30 min. After a further 10 min the end of the kiln is reached, and the dish and contents may be removed whilst still red hot. A kiln designed for the purpose would require exit and entry doors, and the whole process would need to be controlled by microprocessor. Such a furnace need be only about 1.2–1.5 m long and about 0.3 m wide and high, allowing a pair of dishes to pass through the kiln side by side. Such a kiln would have a potential throughput of about 70 samples per day, as against a maximum of 15–20 per small laboratory furnace.

Fig. 2.9. Photograph of prototype electric tunnel kiln for losses on ignition

2.7 PLATINUM WARE

Most of the 'platinum' ware used in an XRF laboratory will probably be platinum/gold alloy of composition 95% Pt/5% Au. Currently, this is almost an universally used alloy, as it is the best non-wetting material available. Various minor modifications have been made to it with the intention of increasing its useful life. Rhodium has been added, also zirconia and even yttria. The rhodium addition was made to increase the hardness and the others to inhibit crystal growth, either of which would help the ware to survive more fusions. Samples of all these types have been tried; the grain-stabilized type do appear to last longer provided the danger of premature failure from use with unsuitable samples can be avoided. Premature failure is always a risk for a laboratory analysing samples without control of their nature or composition. This makes grain-stabilized ware particularly suitable for casting moulds, since at this stage of the procedure the risk of adverse reaction to the sample has passed.

Both the fusion dish and the casting mould tend to have hard lives, being used in adverse conditions. Both are subject to rapid temperature changes, and the fusion dish has to withstand prolonged heating and the attack of a very corrosive flux. The casting mould is subject to rapid heating and cooling cycles, the latter under the stresses created by the solidifying melt. Casts tend to create slight distortion of the base of the mould. This, together with the slight roughening of the metal surface of the mould, can be accommodated for some time by polishing the upper surface of the base of the mould with materials such as jewellers' rouge. Eventually, however, the curvature can only be corrected by re-pressing the mould in a die or returning it to the manufacturers for replacement.

2.7.1 Size and Shape of Ware

The parameter that decides the size of the platinum/gold ware used is the diameter of the bead. A variety of techniques involving bead diameters ranging from 25 to 40 mm is in use. The specific procedure with which this book is primarily concerned, and for which detailed dimensions are given, uses a 35 mm diameter bead. The thickness of the bead needs to exceed the critical depth for the element lines used in the analysis, and this and the diameter of the bead represent the determining factors. These determine the volume of melt necessary to fill the casting mould, which in turn determines the weight of flux/sample mix and so the volume of the fusion dish. The use of a 35 mm bead requires the use of a 50 ml capacity fusion dish. The volume of sample and powdered flux required for this size of bead cannot be easily mixed in a smaller vessel without the risk of spillage.

The casting mould is a shallow, flat bottomed circular cup with straight angled sides leading to an upper circular or square surface. The angles and the radii of curvature which create them at the bottom and the top should be such that the bead will easily release. If the upper surface is to be used, the upper angle,

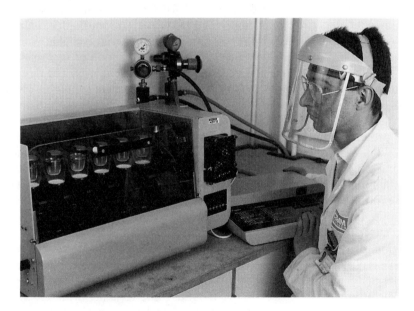

Fig. 2.10. Photograph of Claisse automatic fusion equipment

APPARATUS AND EQUIPMENT

Fig. 2.11. Photograph of prototype combined fusion/casting dish

although not holding the bead, should lead the melt to take up an angle of rest so as to produce convex beads. The upper plane can take the form of a narrow circular flange, but from the point of view of safe handling it is better for manual casting to use moulds that have been pressed out of a square piece of metal, leaving four possible tags to be gripped by the tongs. Platinum can be saved by having a single square flange on an otherwise circular shape.

Many automatic fusion apparatuses (Fig. 2.10) are available, and each calls for its own shape and size of platinum ware. Purchase of this type of apparatus has to be considered in the light of compatibility with current practice and equipment. Wholesale changes to calibrations and replacement of platinum currently in use have to be taken into account when assessing potential costs and savings from automation.

Present automation equipment often requires the bead to be transferred from the fusion dish into a casting mould by mechanical pouring, thereby greatly increasing the mechanical and electronic complexity, and hence the cost. Alternatively, the bead is allowed to cool in the fusion vessel (of conventional shape, but with a flat rather than concave base), resulting in a very variable and more importantly, concave upper surface. A combined fusion and casting dish (Fig. 2.11), as investigated by Ceram Research, to enable the upper surface of the bead to be used, would include what is effectively a casting mould, as the base of a fusion dish with the sides of a fusion dish rising from the upper flange of the casting mould. This has been used successfully for fusing and casting beads, but the main problem appears to be the cost arising both from heavy manufacturing charges and the weight of the platinum needed.

2.7.2 Care and Maintenance

Alkaline fluxes attack platinum, a fact that is demonstrated by successive weighings of the dish after fusions; a loss in weight of a few mg is normal.

The process of fusion, with its rapid changes of temperature, and the need to handle the platinum with tongs create some deformation. At high temperatures platinum is even easier to distort than when cold. It is therefore advisable to re-shape the vessel between each use using specially designed formers. These can be obtained from the platinum suppliers and, although expensive, are essential. Male and female formers are available; it is best to purchase and re-form the vessel between the two halves.

Platinum ware, when hot, should always be handled with platinum tipped tongs, otherwise damage may arise from the formation of an alloy at the point of contact. Generally, these tongs will be made of nickel or stainless steel; the latter appear currently to be the more popular, or at least the more readily available. Nickel might, in fact, be better, as the metal of the tongs becomes hot in service and nickel is less likely to oxidize under these conditions. Oxidation can result in two defects. First, the conversion of metal to the oxide is accompanied by an increase in volume, which causes the platinum tips to burst. Also, just above the tips, the metal appears to burn away, causing 'necking' and ultimate breaking away of the tip. Unfortunately, most nickel tongs are joined with a simple rivet which, after some use, may stretch, giving rise to poor grip and, hence, danger from hot melts. Stainless steel tongs can be bought as a boxed type, which do not have this dangerous tendency.

After each use it is necessary to ensure complete cleanliness of the vessel before re-use. Some small amount of melt may well be retained in the dish. Soaking in acid can remove it in time, but a better practice is to add a small amount of sodium carbonate and a little boric acid, and to melt this mixture with the remaining melt over a gas burner, swirling to ensure treatment of the whole of the internal surface. This should be continued for one or two minutes to ensure incorporation of all the residue into the cleaning flux. The proportions of sodium carbonate and boric acid should be so arranged that there is a considerable excess of the former, so that, when the melt is brought into contact with acid, dissolution will be very rapid and the liquid will effectively be stirred by the evolution of carbon dioxide. The melt may be allowed to cool and then placed into a 250 ml glass beaker containing about 80–100 ml of diluted hydrochloric acid (say $1+3$). If there has been any evidence of the presence of iron alloyed with the platinum it is necessary to stew the vessel in hot hydrochloric acid $(1+1)$*, preferably overnight or in an ultrasonic bath.

---

*Acid concentrations are given $(x+y)$ indicating that x volumes of the concentrated acid is added to y volumes of distilled water.

APPARATUS AND EQUIPMENT   31

Casting moulds should not become seriously contaminated in use, and a brief soak in hot diluted hydrochloric acid (1 + 1), followed by rinsing with distilled water and drying by heating, should make them available for re-use. After casting samples that contain problem elements, such as those that tend to stick, it is safer to stand the moulds in the hot acid overnight.

Damage can occur to platinum during both the determination of loss on ignition and decomposition processes. The danger arising from carbon and carbonaceous matter producing carbides and alloys with reduced iron is covered elsewhere in some detail. Problems arising from heavy metals etc. will be referred to more specifically under the sections on individual sample types. The presence of silicon in reduced form, particularly as the metal or carbide, is a specific problem which will also be the subject of further comment. The one other main problem occasionally met with in some unusual samples is the presence of constituents in the metallic form. This can occur with elements such as silver, gold or similar platinum type metals being added to the material as such, or when dealing with used materials, catalysts or slags that have been in contact with elemental metals such as iron, aluminium, copper etc. The latter group cause serious problems if they are not identified beforehand; they can generally only be successfully fused by a preliminary oxidation to the oxide (e.g. by evaporation with nitric acid) and then analysed as total oxide. The quantification of the metal content has to be left, if possible, to other techniques.

2.8 SELECTION OF SPECTROMETER

The range of types of X-ray spectrometers currently available is broad. The sequential type of instrument based on a simple goniometer with manual settings of all parameters has effectively disappeared, as instruments are now normally offered with total microprocessor control in a variety of versions with increasing levels of automation. Simultaneous spectrometers have changed less in principle, but modern versions can offer a level of sophistication beyond imagination a decade or so ago. In addition, increasing use is being made of simultaneous/sequential instruments which combine many of the advantages of both. In all cases the built-in computer usually includes data storage, both of results and of the information required for conversion of raw data into computed analyses. Computer memory is also used to retain a series of programme parameters pre-set for the analysis of ranges of sample types and often previous results, specification details, quality control data etc., to which reference can be made to check currently achieved figures.

Energy dispersive X-ray spectrometers are not appropriate to the types of accurate analysis that form the subject of this book. They are very useful for a wide range of analytical activities, but are not within the range to be considered for the analysis of materials where major contents of the oxides of light elements, such as magnesium, aluminium and silicon, or minor amounts of sodium need to be determined to a high degree of accuracy.

2.8.1 Type of Spectrometer

It is unlikely that a purely manual or semi-manual instrument would even be considered, except as a second-hand purchase by a relatively small laboratory. Speed of analysis is much slower and cost per analysis much higher than with any other type.

A simultaneous instrument (Fig 2.12) has the very great merit of simplicity in use. Once the spectrometer is set up, samples can be processed with comparatively little attention. Each channel has its own separate crystal, collimator and detector, and in consequence there are many components that can bring about a stoppage. Its inherent weakness is that only those elements for which there are fixed channels can be determined, limiting the range of materials that can be fully analysed. On the other hand there are few mechanical parts to fail, and even with modern technology this can be a major advantage in reducing 'down-time'.

A sequential spectrometer (Fig. 2.13) needs a vacuum path goniometer for work with materials of the type being considered, as many of the determinands are light elements. Parameters can be pre-set to give programmes for a wide range of materials. As samples become more complex and involve an increasing number of determinations, a number of which may need

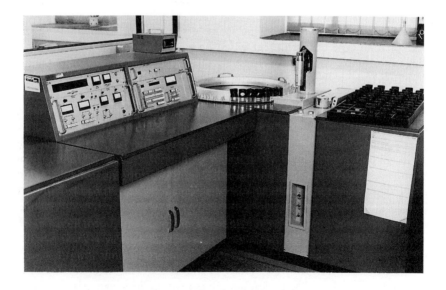

Fig. 2.12. Photograph of Telsec TXRF X-ray spectrometer

APPARATUS AND EQUIPMENT 33

Fig. 2.13. Photograph of ARL 8480 X-ray spectrometer

relatively long counting times, the total time needed for an analysis can become over-prolonged. The use of moving parts to set up the various parameters introduces both the possibility of mechanical failure and slight tolerances at each setting.

A compromise can be achieved by using a simultaneous/sequential instrument (Fig. 2.14). This combines the virtues of both previous types. Wise selection of fixed channels to cover long counting times and the most frequently required elements, together with the use of the goniometer for the remainder, can produce complex analyses quickly. If the sequential system involves a goniometer in an air-path mode then, clearly, all elements demanding a vacuum path have to be on fixed channels. If a simultaneous/sequential spectrometer is to be the choice, it has to be remembered that some of the disadvantages of each type will have to be accepted. The goniometer will have its mechanical parts and the fixed channels will possess the extra components. The decision as to which type to buy will be influenced by cost and the range of sample types to be analysed.

The addition of a sample changer will enhance throughput considerably and will permit unattended night operation. Apart from large analytical programmes using a sequential spectrometer, night operation will almost certainly be needed if the laboratory is handling 'unknown' samples that require a preliminary qualitative or semi-quantitative analysis. For this type of work, a goniometer type spectrometer is an obvious requirement unless alternative techniques are

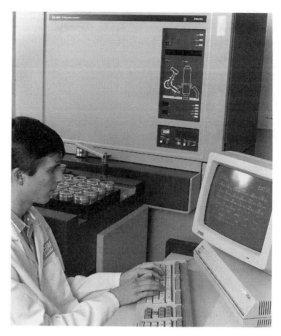

Fig. 2.14. Photograph of Philips PW 1606 X-ray spectrometer

available. Sample changers with less than 50–70 stations are not very cost-effective.

Fused, cast beads tend to be smaller than the X-ray beam, so that part of the sample-cup mask may be irradiated. One possibility is to have these masks made of an element not often determined, e.g. gold, silver or rhodium. Rhodium would be ideal, as many X-ray tubes use it as target, but it is expensive to work. Silver has the advantage over gold in that it interferes with fewer lines. Unfortunately masks can become contaminated in use. This can be a serious problem when the side of the mask nearest to the X-ray tube is contaminated, especially with silica (from vacuum grease) or sulphur (particularly in the case of silver masks), so that the relevant channels also need masking.

A sequential instrument will generally have some limitation with regard to the number of detectors (normally one, two or three), collimators (normally two), or crystals (five or six), unless twin goniometers are fitted. Collimators are normally set by the manufacturers, coarse for light elements and fine for heavy. Flow counter and scintillation detectors are probably the best choice if only two detectors are to be employed, with the addition of a sealed counter (preferably krypton) if a third is possible.

APPARATUS AND EQUIPMENT

It is suggested that currently the best selection of crystals for a special instrument is:

TlAP	for Mg, Na, F and O
PE	for Al and Si
Ge	for P, S, Pb $M\alpha$, Cl, Zr, $L\alpha$, etc.
LiF 200	for most determinations
LiF 220	for line overlap problems

If a sixth crystal is possible, laboratory needs may well indicate the value of InSb for Si $K\alpha$, Y $L\alpha$ or Sr $L\alpha$, or possibly a layered pseudo-crystal, especially for light element work (Mg, Na, F, O and C).

One important factor in the use of sample holders, and therefore changers, is to ensure that each cup yields the same intensities with the same sample. This needs to be checked at an early stage of the acceptance trials after installation. Consistency is achieved only by accurate machining of the cups and the sample masks; this makes it important that the cup and mask are used as a unit and not separated, otherwise intensities may be affected. The Telsec TXRF avoids this problem with the cups because the reference position of the bead is set by the spectrometer, the beads resting on the internal mask, not the cup, during measurement.

2.9 COMPUTER

2.9.1 Computer Hardware

Experience in many fields has indicated that, whatever size of computer hardware appears correct at the time, this capacity should be at least doubled. It is better to choose a wide printer rather than a narrow one, if only to assist in printing clearer graphics. It is possible that a normal size printer may be desirable in addition if high quality reports are required from the system. An adequate buffer store to the printer is needed so that data can be down-loaded from the computer, thus allowing the spectrometer operator to continue inputting data, instructions or results. If other software is run on the computer, or home-written software is developed, a second terminal will prove extremely useful.

2.9.2 Computer Software

There can be no good spectrometer system without good software. This may appear so self evident as not to be worth stating, but problems arise so frequently from this source that it warrants emphasis. Software has often been the weakest point in the package offered by the manufacturers. A common danger, at least in the past, has been that new systems have been refined by the manufacturer using the customer as a guinea-pig, a situation that can prove disastrous.

There is often a yawning gulf between the analytical chemist and the computer programmer. This inability of computer software suppliers and the ultimate users of their work to undersand each others' problems is most certainly not confined to science. It is not easy to bridge the gap between the analyst and the computer expert and, as a result, many systems fail to come up to expectations.

Software supplied by the manufacturers has improved very markedly over the years and it is now usually possible to select an adequate system by careful prior examination. It is, nevertheless, highly desirable that the software purchased should be capable of being modified by the user rather than having to refer back to the supplier every time. The latter course can be very expensive and does not always yield the desired outcome.

It is essential that the computer software gives the analyst all the facilities that are needed. Many of the applications of XRF originated from situations such as the iron and steel industry, where it was not possible to use synthetic standards. As a result many of the systems have tended to operate on the basis of the use of regression analysis, which is not ideal for this sort of work.

In addition to many technical factors which have to be handled, any computer software must allow for the entry of data concerning non-XRF determinable components (Li_2O, F, etc.) and loss on ignition and the use of such data in calculating the final analysis. It should also be possible to make allowance for the contamination by tungsten carbide during grinding. Final output of the data needs to be both in a form suitable for the analyst's inspection and, ultimately, in the format of the report to be issued to the 'client'.

As the situation is developing wherein greater emphasis is being placed on Accreditation and Quality Assurance schemes, software in the future will need to take this factor increasingly into account. Quality control checks and records of this and other required data should be capable of being acquired and stored ready for inspection as required.

2.10 REAGENTS AND BLANK DETERMINATIONS

The problem of blank determinations in XRF analysis is by no means as acute as in 'wet' work. The reason is not far to seek; so few reagents are used. Essentially, the only reagents of any significance are the materials used in the fluxes. These have to be of a high standard of purity, and are usually bought in as a special grade made expressly for the purpose. It has to be remembered that there is present in the bead used for analysis five times the weight of flux relative to that of sample. Thus, an impurity equivalent to 0.01% by weight in the flux appears as an error of 0.05% in the result for the sample.

Experience has shown that even XRF grades of reagent cannot be totally trusted. Great care is taken in their preparation, but nevertheless it still behoves

APPARATUS AND EQUIPMENT

the analyst to maintain a constant system of checking and control. Impurities such as silica, lime, magnesia, sulphur and phosphorus are the most commonly found.

When first calibrating, it is desirable to ensure that levels of impurity are low, so that the counts obtained for each element can be used as a norm for the future. New batches of flux are then checked against this norm and deviations measured. Minor or negative deviations can be accepted and allowed for by replacing zero ratio standards and in computing the results. If the change is both positive and higher than is considered satisfactory the batch of flux is best rejected. It should be noted that, in principle, if the impurity is evenly spread throughout the flux, and the calibration and interference measurements are made on the same batch as the analyses, no error will ensue, as 'blanks' are automatically accounted for in the computation. Again, given even distribution, any change from batch to batch, if reasonable in size, can be tolerated and corrected for by simple addition or subtraction from the result. Infrequently it is possible to have batches of flux where impurities are not evenly distributed. In the worst case they can vary not only from container to container but even from fusion to fusion; this is obviously not acceptable.

Since the work entailed in introducing a new batch of flux can be considerable, it is advisable to do this as infrequently as possible. Thus, batch sizes of flux should be as high as is practicable and fresh batches should be obtained in good time so as to be able to test them thoroughly sufficiently long before the previous batch is exhausted to allow for the possibility of rejection.

3 Determination of Non-XRF Elements

3.1 INTRODUCTION

XRF, although probably the most universally useful tool in a laboratory handling the types of materials dealt with in this book, is incapable of being used at all for determining some elements. Furthermore, its use with the fused, cast bead technique increases the number of such elements. Nor is it the most generally suitable technique for the determination of trace elements. It needs to be complemented in the laboratory by other suitable techniques. This chapter discusses some of the limitations of XRF and ways in which the gaps can be plugged.

There are several reasons why elements cannot be determined by XRF. First there are the light elements, say, those with atomic numbers below that of sodium that do not respond sufficiently to X-radiation to allow their determination to an adequate level of precision. Improvements in instrumentation are reducing the number of these; for example, it now might be possible to include fluorine as a determinable element but, as yet, hardly fully quantitatively. Lithium and boron are also among these light elements, which is one of the main reasons why lithium borates make ideal fluxes. These fluxes allow the ready penetration of X-rays, so that they yield very suitable beads with low mass-absorption coefficients and, as the elements cannot be determined by XRF, they can be added to the melt with no loss to the analyst. Also 'lost' to XRF are hydrogen, carbon, nitrogen and oxygen, all of which would, in any event, be lost during fusion. The determination of nitrogen and oxygen is, of course, not generally attempted or requested except in samples of reduced materials such as silicon carbide or nitride. Beryllium is also lost, even though it would rarely be sought. On the few occasions when it might be required it would pose serious problems because of its highly toxic nature. Finally, there is helium.

This book is solely concerned with the fused, cast bead method of analysis, so that it is necessary to consider how this procedure can introduce limitations that prevent determinations. These take, in principle, two forms. Oxides or elements that are partially or wholly volatile during the fusion process are not determinable, neither are elements that are easily reduced, as they then tend to alloy with the metal of the dish (some elements, e.g. silver, may well be present in some samples in the metallic state).

Elements such as fluorine and the other halogens are not easy to retain during the fusion unless the matrix is favourable, i.e. alkaline in nature and relatively free from silica. The volatility (during fusion) of both silicon tetrafluoride and boron trifluoride tends to militate against the successful determination of fluorine. Similarly, sulphur as sulphide or sulphite will be lost unless great care is taken to ensure its early oxidation to sulphate. Even sulphate will be lost except in favourable matrices; even so, its retention requires lower fusion temperatures and the right degree of care. Sulphate can, in practice, often be determined with sufficient accuracy for many purposes, but it will sometimes be necessary to turn to other techniques. Selenium, amongst the less commonly required constituents, occurs almost invariably in ceramic materials as selenide, and is therefore almost totally lost, so that recourse needs to be made to other approaches.

Apart from these anionic radicals, a number of the metallic oxides can give rise to problems. Volatile oxides, such as those of arsenic, antimony, zinc, lead, vanadium and cadmium, may cause difficulties to various degrees, requiring lower fusion temperatures (1050°C) and care; their determination is, however, generally possible.

A number of the above elements also give rise to problems in that they are amongst those elements with easily reducible oxides. Lead and zinc are good examples. They are joined by elements such as copper, cobalt and even iron. Even so, these elements are normally determinable with care, with the proviso that in the case of the copper and cobalt it may be necessary to limit their content in the glass of the bead by extra dilution. Finally in this class are the noble metals. On the infrequent occasions when they are present, they are usually in the form of the element, and as such quickly react to join the alloy of the dish. Gold, platinum and most levels of silver invariably require the use of other techniques, although silver at low levels may be determinable under suitable conditions.

These problems are dealt with in more detail elsewhere in the book; the concern here is to draw attention to the need for other methods of analysis to be available. It is not intended that full procedures using these other techniques should be given, as this does not lie within the remit of this volume and would occupy too much space. Most of the methods referred to are known to analysts in this field, and references are therefore not given to all the procedures mentioned. The most appropriate methods for any given laboratory remain, as will be emphasized on a number of occasions throughout the book, dependent on the volume and type of work to be carried out. For example, in any specific case the choice between techniques, say, wet methods, AAS and ICP, may well be determined by the potential frequency of the determinations involved and the relative difficulty of the procedures. In this assessment it has to be borne in mind that instrumental methods involve originally high capital cost, followed by maintenance and running costs. They also demand costly setting up and maintenance of calibrations.

Wherever possible, traditional 'wet' methods are best avoided. They are time-consuming, both in terms of analyst time and elapsed time. They also demand skill and experience for their successful completion, more so than most instrumental methods once the latter have been developed. In terms of labour they are almost always considerably more expensive per test. There are occasions where wet methods are unavoidable, but these are decreasing with the passage of time.

3.2 METHODS FOR NON-XRF ELEMENTS

3.2.1 Determination of Borate

There are two traditional methods worth considering here. The first is that generally employed, particularly in the glass industry, for this determination.[1] This involves fusion of the sample and decomposition of the melt in water. Interfering elements are removed by precipitation with calcium carbonate and, after filtration, the solution is titrated with standard sodium hydroxide to neutrality. Finally, mannitol or sorbitol is added to form the stronger mannito- or sorbito-boric acid and this is then titrated with sodium hydroxide. With careful manipulation good results are obtainable, but it is not an easy method to use and can yield 'sport' results. In good hands it is claimed to produce results with an accuracy of ±0.15% absolute, and two or three determinations per analyst per day in duplicate appear to be about the norm. It is subject to a few interferences that need the application of modifications.

An alternative procedure involving separation of the borate via distillation with methyl alcohol lost favour many years ago and is rarely used. A further method offering apparent simplicity is pyrohydrolysis, wherein the borate is removed by passing steam over the sample at about 1400°C and condensing the distillate, then determined using a similar titration with sodium hydroxide as above. The snag is that platinum apparatus is generally required, involving very heavy capital costs in setting up and maintenance, so that the method is not in general use.

A more recently adopted method involves a colorimetric finish[2] that requires less manipulation, thereby allowing roughly twice the number of determinations per analyst per day as the double titration method. After fusion, an aliquot of the solution is reacted with carminic acid and the colour is measured. This was originally used for the measurement of low levels of boron, such as in steels and magnesites, but has proved successful for contents up to about 25% of B_2O_3 in glasses, glazes[2a] etc. It appears capable of producing accuracies of about ±0.2–0.25% absolute with greater ease than the volumetric method.

It has the disadvantage that the reaction and measurement are carried out in virtually concentrated sulphuric acid (as are almost all the colorimetric methods for boron). In view of the very high levels of dilution and, hence, the high multiplication factors, good manipulative care is needed. Fluorine and titanium interfere but the latter can easily be allowed for. The procedure has been used by the authors for a number of years with relatively few problems.

Boron is usually required at one of two levels of content. Trace element levels are important when the material is to be used in connection with the atomic energy industry or when magnesia or magnesite is to be used as a refractory. It is also required as a minor or major component, up to about 25% B_2O_3 content in glasses, glazes etc. Until recently it has been a problem element, difficult to determine by traditional methods (as suggested above), and there was no obvious instrumental method in sight. There have been techniques based on neutron activation but the processes are not generally suitable for run-of-the-mill laboratory work, being chiefly of value for low levels of content in the absence of other neutron absorbers, AAS is very insensitive even for the determination of major contents, and the use of furnace excitation does not overcome this. Optical spectography is only really suitable when applied to the original sample, and gives rise to spectrographic levels of precision which are totally inadequate for this type of work. However, in the last two years or so it has proved possible to attain adequate precisions and accuracies using ICP spectrometry. These have been achieved using a sequential instrument, so that even better figures should be possible using a simultaneous type.

3.2.2 Determination of Sulphur

There are several methods for the determination of low levels of sulphur by purely traditional methods; the one most recently attracting attention involves reducing the sulphur compounds to the sulphide and measuring the latter colorimetrically. The traditional method for the determination of sulphur[3] at minor or major content levels has been, and for the most part still is, that of a gravimetric finish by weighing barium sulphate. This normally involves a fusion, decomposition of the melt in water and the removal of interfering ions by filtration. After acidifying with hydrochloric acid, the barium sulphate is normally simply precipitated with barium chloride solution, and the precipitate is filtered off and weighed. Some analysts prefer to remove other possible interferents by an additional precipitation with ammonia; in practice the errors introduced by the additional manipulation could well counterbalance the possible removal of interference effects.

This gravimetric method is still almost the only procedure for high levels of content where maximum accuracies are needed; nevertheless it requires careful handling if significant errors are to be avoided. In addition, it may possibly require a double fusion, first of the sample and then of its residue after extraction with water. A fusion and re-precipitation of the barium sulphate is also recommended in some books. If all these processes are followed in attempting to achieve maximum accuracy the method becomes both time-consuming and complex. In many cases the additional refinements could prove counter-productive, with the extra manipulation introducing as many errors as it overcomes. In practice a careful fusion using suitable conditions of time and temperature followed by an XRF fused, cast bead finish can often be virtually as reliable as using a barium sulphate precipitate for such materials as calcium sulphate.

Most sulphur determinations of up to 2-3% content are achievable using combustion methods. There is a simple technique in which the sulphur, in the sample mixed with benzoic acid, is combusted in a sealed glass vessel filled with oxygen, the sulphate produced being determined by titration.[4] The method is of quite adequate accuracy for routine purposes and is fairly rapid, the whole determination being completed in a matter of minutes.

However, where a number of determinations are liable to be required it is probably more satisfactory to employ a furnace combustion method, preferably using an induction furnace, such as are commercially available. Depending on the throughput, the equipment can be such that the finish is manual or semi-manual, by titration, or by measuring the sulphur gases with an infra red spectrometer, the cost of the machine being much greater if the latter system is used.

3.2.3 Determination of Carbon

Carbon, like fluorine, is below the normal limit of detection of XRF and needs to be determined in other ways. Until relatively recently this has been done almost exclusively using traditional combustion methods based on the BS Standard Method for the determination of carbon in coal,[5] involving careful control of temperatures in various parts of the tube furnace together with a range of reagents packed in specific areas of the tube to remove other gases that could interfere with a gravimetric determination of the released CO_2. Correction for carbonate after heating to these relatively low temperatures is always problematical as it is questionable how much, if any, of the various carbonates decomposes. The wide variety of matrices met with in this sort of work makes this evaluation even more complex.

More recently, induction furnace combustion apparatuses similar to that used for the determination of sulphur are being increasingly used. If the determination is to be finished manually, the carbon dioxide is measured gravimetrically after passing through a chain of absorbents. With an infra-red finish the absorption due to the carbon dioxide is measured directly. This can be done simultaneously with the measurement of the absorption due to sulphur dioxide, thus permitting the two determinations to be accomplished in parallel.

Carbon present as carbonate (or bicarbonate) is usually determined by evolution of carbon dioxide with acid; phosphoric acid is normally recommended for this purpose. Weighing the evolved carbon dioxide after absorption by, say, Carbosorb is often the easiest finish.

3.2.4 Determination of Lithia

The determination of lithia is rarely attempted by wet methods as the separations from other alkalis are extremely complicated and difficult. In fact, prior to the availability of simple, inexpensive filter-type flame photometers, lithia was rarely determined except in lithium-bearing minerals. To include the determination of lithia with the flame-photometric determination of the other alkalis is very simple and adds only a few minutes to the time; for this reason its determination became normal. Alternatively, both AAS and ICP can be used, but these more expensive techniques are rarely justified unless the apparatus is needed for other purposes and can be used for other determinations on the same sample solution.

3.2.5 Determination of Halogens

The determination of fluorine by traditional methods calls for the use of very time-consuming and difficult procedures. In fact, these types of methods have been avoided whenever possible.

A relatively simple technique involves the use of pyrohydrolysis,[6] namely heating the sample in a tube whilst passing a current of steam over it. The sample is mixed with a 'catalyst' such as vanadium pentoxide or tungstic oxide. The vapour is condensed and the fluoride in the solution is measured by one of a number of possible methods, of which the use of a fluoride electrode with a pH meter is probably the best and most frequently employed.

The remaining halogens, if present singly, are easily determined by a silver nitrate precipitation. Unfortunately, with the possible exception of chloride, this is not a condition often encountered. It is frequently possible to assume that chloride is present on its own, and for this assumption to be sufficiently correct for all practical purposes. Where more than one of chloride, bromide and iodide are present, traditional methods are best forgotten unless there is no other option, as recourse has to be made to doubtful separatory methods.

Quite recently a solution to this problem and an alternative method for fluorine has become available. This is ion chromatography, which seems to offer solutions not only to this but also to the determination of mixtures of other anions (and some cations) such as may arise in this sort of work.

3.2.6 Determination of Selenium

Selenium is readily volatile, so that procedures for bringing it into solution using a fusion technique are fraught with problems. Similarly, the generally recommended methods for acid decomposition using nitric acid both as an acid and as an oxidant to convert selenide to the more stable selenate have also to be viewed with suspicion. In most ceramic usage selenium is present as a simple selenide or a sulphoselenide, so that it is a moot point whether conditions can be achieved whereby conversion to selenate occurs before selenium is lost from the solution as hydrogen selenide.

Certainly, as is discussed elsewhere, experience has shown that analytical results are almost always very significantly lower than would be expected from the amount added during manufacture. This appears to remain true irrespective of the method of attack and the finish; results by radioactive methods tend to agree with the analytical findings by these more traditional procedures. It is suspected that manufacturing losses could account for the great majority of the selenium added, and that the colour obtained is produced by much less selenium than has been thought. 'Wet' methods of determination are not very satisfactory and recourse has to be made to instrumental procedures.

Both AAS and ICP give satisfactory sensitivities and results for selenium, provided that the problem of getting the element totally into solution can be solved. Even so it is necessary to use the hydride generation technique, that is, reducing selenium to hydrogen selenide with, say, sodium borohydride and then sweeping this into the flame or plasma with a stream of inert gas.

3.2.7 Determination of Volatile or Alloying Elements

This section is kept to generalities, since a discussion of all the methods applicable to this whole range of oxides and covering the range from 0.01% to major contents is beyond the scope of this book. With care, most of the oxides mentioned as volatile or tending to alloy (ignoring for the moment the noble metals) can often be determined by XRF. Cadmium, arsenic and antimony are not easily determinable by traditional methods. Problems are almost sure to arise in attaining clean separations prior to the analytical finishes, but by the use of instrumental methods these difficulties can be avoided.

Almost all the elements in these categories can be determined either by AAS or ICP, although some of them may be difficult to get into solution without loss caused by either the application of heat or hydrofluoric acid treatment. It is usually possible, however, to avoid these difficulties by using lower temperatures of fusion or suitable additions during the hydrofluoric acid treatment and subsequent evaporation. The use of either spectrometric technique requires thorough investigation of interferences and proper method development, as was the case for all traditional methods. The apparent simplicity of the techniques can lull the analyst into the assumption that responses can be taken at their face value. Frequent experiences have demonstrated that this is not so. For those elements amenable to it, hydride generation can often yield better sensitivities and offer less chance of line interference, but with the disadvantage of much poorer reproducibilities and possible chemical interference (iron being a frequent culprit).

REFERENCES

1. Bennett, H. and Reed, R. A. (1971) *Chemical Methods of Silicate Analysis*, p. 165, Academic Press, London and New York.
2. Shelton, N. F. C. and Reed, R. A. (1976) *Analyst*, **101**, 396.
2a. Reed, R. A. (1977) *Analyst*, **102**, 831.
3. Bennett, H. and Reed, R. A. (1971) *Chemical Methods of Silicate Analysis*, p. 205, Academic Press, London and New York.
4. Beech, D. G. (1974) *Testing Methods for Brick and Tile Manufacture*, Special Publication No. 64, British Ceramic Research Ltd, Stoke-on-Trent.
5. *Methods for the Analysis and Testing of Coal and Coke* (1958) BS 1016: Part 6.
6. Bennett, H. and Reed, R. A. (1971) *Chemical Methods of Silicate Analysis*, p. 211, Academic Press, London and New York.

4 Loss on Ignition

4.1 INTRODUCTION

In ceramic analysis the determination of loss on ignition is unique in that it does not measure a percentage *content*. Also, it is carried out following a totally empirical, albeit specified, procedure. In other words, the resulting figure does not represent an absolute content in the material being analysed such as are provided by the oxide or element contents. The determination is much more similar to such methods of chemical testing as measuring the degree of resistance to acid attack than to a true determination of content. Nevertheless, it is a very useful and convenient determination provided that its limitations are appreciated.

Provided that the analyst is dealing with the simple range of materials used in normal bulk ceramic manufacture, or the geochemical analyst has no interest in the individual gaseous components lost during ignition to the selected temperature, the determination of a loss on ignition can provide useful information in its own right. It can also form the final link in the chain leading to an analytical total, i.e. the summation of the figures from all the determinations. This, everything else being equal, should approximate to 100%, and if this is so the analyst can have considerable confidence in the individual figures obtained. Thus, given the specificity of XRF, this analytical total, if satisfactory, can often save the need for duplicate analyses, at least, for routine work.

Confidence in the analytical total can never be absolute, experience having demonstrated only too frequently that the long arm of coincidence can reach out and suggest that erroneous results are apparently correct. This was more to be expected using older chemical methods of analysis, where determinations were much less specific than those obtained by XRF. Some oxides were not determined at all but their content could still affect the results by interfering, usually positively, in one or more of the other determinations. The specificity of XRF greatly reduces the frequency of this sort of coincidence, but the analyst must always be alert to the possibility of the presence of other elements liable to yield line overlaps and/or inter-element effects capable of giving coincidentally balancing errors. Unlikely though it may be, such errors can and do occur. Small but significant amounts of one element can sometimes almost completely compensate for errors on others, to within the levels of accuracy capable of being revealed by an analytical total. Further, many glazes and glasses can contain additions of lithia, although the amount is usually small when expressed as a percentage content because of its low equivalent weight. It is very easy to 'lose', say, 0.25% of Li_2O, because XRF does not pick it up and because it is not enough to lower the analytical total to below tolerance levels. Similar problems can arise with small contents of fluorine or boric acid and, less frequently, with beryllia.

Even the geological analyst, using XRF, will need to consider seriously the use of ignited samples because of the need to maintain the ratio of flux to sample in the final bead. Thus, whether or not an actual figure for the loss on ignition is obtained and used, the process of igniting the sample will still need to be carried through. Thus the reactions encountered during the ignition will have relevance.

The determination of loss on ignition is carried out by progressively raising the temperature of a portion of the sample [usually the dried (110°C) material], traditionally 1 g, but with XRF analysis more usually 2 g, to a specified temperature, commonly 1025°C, and then maintaining that temperature for a specified length of time, usually 30 min. The difference in weight between the original portion of the sample and that remaining after the ignition, expressed as a percentage of the original weight, is the 'loss on ignition'. In theory, the ignition is repeated to constant weight, but experience shows that for most common materials heating for 30 minutes is ample.

It may well be thought that 1025°C is a strange temperature to select for this determination, and it would be hard to disagree. It arises as one of the peculiarities in the development of international standard methods of analysis, a topic returned to later in the book. In the UK the accepted temperature for the determination of loss on ignition was 1000°C, whereas on the European continent it was 1050°C. Neither temperature has the advantage of being 'correct' so that the obvious way in which a compromise acceptable to all parties could be achieved was to permit a temperature of $1025 \pm 25°C$. A judgement of Solomon, perhaps?

The above is the normally prescribed procedure used in standard methods for most ceramic materials, but there are exceptions. Some of these are discussed later, together with the limitations of the determination itself when applied to materials of complex and complicated composition. However, it is necessary to discuss here the nature of the usual reactions taking place during the heating process that result in changes in weight, the sum total of which will be reported as the loss on ignition.

The reason for measuring the loss on ignition is primarily that the simple procedure provides information adequate to the needs of the ceramic technologist. With such information it is possible to evaluate the raw materials

LOSS ON IGNITION

and calculate both the mix composition required to give the desired product and the composition of that product. The determination is much faster and cheaper to carry out than the several determinations needed to provide the equivalent information. It is therefore normally regarded as the best commercial choice. This does not preclude the analysis of the material for some of the volatile constituents themselves when a knowledge of their contents is of value but, even so, in the field of ceramics this would usually be regarded as additional to the determination of the loss on ignition.

In terms of ceramics 1025°C is a low temperature; in the refractories industry it would be considered very low, as temperatures of use of many refractories are often over 1500°C. Thus, the loss on ignition process does not necessarily bring to completion all the chemical reactions that could occur during firing or use of the product. It is very likely that subsequent heating at higher temperatures will result in further reactions and weight loss. It is important, therefore, for the analyst to know just what data the ceramist needs. There may well be instances where it is desirable, in effect, to attempt to reproduce the production firing conditions so that the user of the data can be made aware of the anticipated composition of the final product. On the other hand, there will almost certainly be occasions when a temperature of 1025°C is too high. In the analysis of glasses, for example, a much lower temperature is advocated, since the material may well melt early in the heating process, inhibiting further loss of volatiles. The same factor frequently applies to the analysis of pottery glazes, but here problems can arise as additions are often made to the frits in the grinding mill to provide the final formulation of the glaze. China clay and calcium carbonate are frequent additives, and a low temperature, for example about 600°C, will fail to decompose either of these completely, giving low analytical totals. In the end, in this instance, the higher temperature may need to be chosen to ensure that the desired decompositions have been completed, so as to make sense of the ultimate set of figures forming the analysis.

From the foregoing, it is clear that the analyst has to be fully aware of both the nature and the likely chemical behaviour of the materials being analysed. One of the most common problems in analysis is the gap that can, and often does, exist in the areas of general chemical knowledge and expertise between the analyst and the technologist. It has often been suggested that it is the place of the analyst to bridge this gap, not that of the technologist.

The causes of the changes of weight that occur during heating depend, obviously, on the nature of the material. Generally speaking, however, the range of types of chemical reaction occurring is limited. Nevertheless, as the loss on ignition is the summation of all the changes in weight caused by the various chemical reactions engendered by the heating process, it seems necessary to discuss them in some detail.

4.2 CHEMICAL CHANGES DUE TO HEATING

One of the most important considerations to be borne in mind when analysing geological materials (and thus, by definition, the vast majority of ceramic materials) is that they are not pure compounds. They are sampled or mined as nature deposited and weathered them. They are then possibly beneficiated, selected or purified and are then used as geological specimens or ceramic raw materials. A few materials are chemically prepared and can be regarded as chemicals as much as raw materials, a sort of inorganic intermediate. Typical of this type are alumina, now finding wide-scale use as a ceramic raw material, and zirconia, whose use as a refractory for special situations and, indeed, as an engineering ceramic shows every sign of growing to a position of major significance.

The distinction between pure chemical compounds and such natural materials is very important in many ways, but here we are concerned solely with the reactions produced by heating and the temperatures at which they are likely to take place. The analyst dealing with a pure chemical compound will generally find, in one of a variety of reference books, details of melting points and/or temperatures of decomposition. The presence of impurities can radically alter the compound's behaviour. Although changes merely involving alteration of the oxidation state may be relatively little affected, decompositions, particularly those entailing the breakdown of carbonates, sulphates etc., will probably occur at temperatures very different from, and usually lower than, those quoted in the literature.

The materials being analysed may vary from almost pure materials with very small levels of a few impurities to complex natural or artificial mixtures. In the first case melting points and decompositions are likely to be affected as would those of a normal chemical compound. The second type of material, naturally enough, will behave like a mixture of compounds. With these there are possible reactions that can occur at elevated temperatures. Obviously, if the mixture were to contain, say, calcium carbonate and silica, one might expect a reaction to produce calcium silicate and carbon dioxide. This reaction will occur at a much lower temperature than that at which the decomposition of calcium carbonate would normally take place. The speed of this and similar reactions will depend on many factors. Among these will be the original composition, the particle sizes of the reactants, the nature of the products and so on. Smaller particle size will increase the intimacy of contact and speed the reaction; a similar effect will occur if one or more of the products is volatile, helping to push the equation from left to right. Similarly, if one of the products is liquid at the temperatures in use there will be a tendency to increase the degree of contact and speed the reaction; in fact, secondary reactions may be started. All these factors tend to produce earlier and faster decompositions and other phenomena than would occur with materials of higher purity.

Most ceramic materials are either naturally very fine or have been processed to yield very low particle sizes/high unit surface areas. Clays are naturally very fine, particularly ball clays. Pottery bodies, glazes, etc. are prepared to a very small particle size, since early and rapid reaction is sought during the manufacturing processes. As a result, the analyst can expect many of the reactions to take place at relatively low temperatures.

Some guidance on this subject can be obtained from data derived from thermal analysis methods, thermogravimetry and differential thermal analysis being the most useful. Even here, however, these procedures use a steadily increasing temperature, so that temperatures at which peaks occur merely indicate the point of maximum speed of the reaction under the particular conditions employed. All the reactions that take place to produce the loss on ignition figure involve not only temperature but time. Once a temperature is reached where a reaction starts to occur, the reaction will almost certainly continue to virtual completion at that temperature, given long enough. In practice, of course, the rate of reaction has to be significant.

In the past, the loss on ignition determination in ceramic analysis rarely presented any serious difficulties, but technical developments in ceramics have resulted in a much broader range of materials to be analysed, some of which have created problems. This situation has been mitigated by the developments in analysis, particularly XRF, which have enabled the analyst to carry out tasks that would otherwise have been technically extremely difficult and commercially impossible. The enormous reduction in the cost of analysis, especially for 'unusual' samples, has brought the analysis of these types of materials within the ambit of the ceramist.

This has meant that the use of the simple device of determining the loss on ignition can result in some difficulties for the analyst. With knowledge of the needs of the ceramist it is possible to decide whether the circumstances call for a straightforward loss on ignition determination, as described in the standard methods, or whether to select specific temperatures and times of ignition, possibly together with such additional determinations as will provide the required data.

Even so, heating, say, 1 g of finely divided material in a platinum dish or crucible with free access of air may allow more complete reaction to take place than would occur in a stack of ware such as plates, tiles or bricks. Thus, the result for the loss on ignition determination has to be interpreted in terms of a knowledge of the materials involved and the processes, particularly firing conditions, to which they will be subjected.

4.3 THE EFFECT OF COMPOSITION

As previously discussed, any deviation from total purity in a chemical compound tends to reduce the temperature at which the reaction occurs. Le Chatelier's principle also applies, so that the less volatile oxide (especially, and usually in these instances, acidic oxides) tends to replace a more volatile one. Thus the more volatile oxide is driven off at a lower temperature than would happen in the pure compound. Most ceramic materials are mixtures of minerals, sometimes with several major components. In many cases, however, the minerals containing relatively volatile acidic oxides, such as carbonate or sulphate, occur as lower-level minor components.

The situation is best illustrated by example. A typical clay used in the manufacture of ceramics may contain, in addition to a clay mineral such as kaolinite, a number of other minerals at different levels of content. Quartz is almost always present, sometimes at very low levels, but more normally ranging from a few per cent up to major amounts. Other aluminosilicates such as micas or feldspars are usually present and provide most of the alkali content. Carbonates are another frequently found impurity, generally in the form of calcium or magnesium carbonates or as dolomite. In clays used by the brick industry the level of calcium carbonate can be quite substantial. Ferrous carbonate, siderite, is another component of brick-making clays, but here the decomposition takes place at a low temperature. Sulphates and sulphides may also be present in various forms, generally as minor impurities only. Most common of the sulphates is calcium sulphate, and the sulphide most commonly found is pyrite, or less frequently, chalcopyrite. It is often assumed that, as minerals decompose in air at relatively low temperatures, this sulphur will burn out during firing and be lost. This is not always true, particularly if the clay has a high lime content, a not uncommon combination. Some at least of the oxidized sulphide may become trapped as calcium sulphate during the firing, as may some of the sulphur in the fuel, if any. Finally, there may be present small amounts of fluoride, chiefly as calcium fluoride, although minerals such as tourmaline may also be present.

Thus, in a clay material there could be present two non-volatile 'acidic' radicals, silicate from the quartz and aluminosilicate from the kaolinite and the fluxes (micas and feldspars). At elevated temperatures one of these, silica, will be relatively highly reactive, the aluminosilicate much less so; that combined with the alkalis will probably be significantly less reactive than the clay based aluminosilicate. There may well also be present small but significant amounts of two volatile acidic oxides, carbon dioxide and sulphur trioxide, in the form of carbonates or sulphates respectively. Lastly, there will be small amounts of fluoride and even other halides, although the latter are usually of less consequence.

On heating, there will be first the commencement of the decomposition of the clay mineral with loss of water; the temperature and speed at which this occurs depend mainly on the nature of the clay mineral, but it can commence even below 100°C. This decomposition will continue progressively as the temperature is increased. If present, minerals such as siderite and pyrites will decompose at these relatively low temperatures and any volatile oils etc. will be driven off. Next there will occur decomposition of magnesium and calcium carbonates (Eqn 4.1) followed by the commencement of the breakdown of calcium sulphate (Eqn 4.2). These decompositions occur at lower temperatures than are quoted in the literature, since the close proximity of a replacement acidic oxide such as silica will enhance the breakdown. Also, at these lower temperatures, carbonaceous matter will be burning out. Finally, there will be some reaction between any calcium fluoride present and silica to produce calcium silicates and silicon tetrafluoride (Eqn 4.3). Because of the ease with which the volatile components can escape, there will be strong tendencies for these reactions to go to virtual completion. Again one is concerned with rates of reaction, and it is clear that intimacy of contact is very important in determining the speed with which the reactions will proceed. As most clays contain significant amounts of very finely divided material, reaction will occur fairly readily. Some of the reactions, however, will be only partial, and it is rare to remove all the sulphate or fluoride at 1025°C.

$$CaCO_3 + SiO_2 = CaSiO_3 + CO_2\uparrow \quad (4.1)$$

$$CaSO_4 + SiO_2 = CaSiO_3 + SO_3\uparrow \quad (4.2)$$

$$2CaF_2 + 3SiO_2 = 2CaSiO_3 + SiF_4\uparrow \quad (4.3)$$

The loss of fluoride will usually involve the loss of silica, and this renders the analysis theoretically in error. These errors normally reach significant proportions only when fluoride is a deliberate addition to a material. This is often the case with vitreous enamels or welding fluxes and, of course, in the steel industry to produce high fluoride-bearing slags. Again, the analyst has to take into account the possible size and relevance of such errors, as well as the cost of acquiring the correct answer by alternative methods as against the degree of need for it. Some mitigation of the potential error often arises because many of the materials with high fluoride contents also contain equivalent amounts of calcium and other metallic oxides, together with a lower level of silica. The materials can thus almost be considered to be alkaline in nature, so that the tendency to lose fluorine is greatly reduced.

Consideration needs always to be given to the anticipated chemical and mineralogical composition when analysing an unfamiliar material, taking note of the balance of acidic and basic oxides; it can frequently be assumed that the aluminosilicates will act amphoterically. Ideally, conditions should be chosen so that reactions are carried to completion, if this is possible, as this will produce a more viable analysis on the ignited basis. The second and less desirable alternative is to attempt not to start reactions that cannot be taken to completion. The least desirable situation is to have one or more reactions only partially completed, but with more complex forms of ceramic this is the most frequently occurring situation.

The complex nature of the reactions that add up to the figure quoted as 'loss on ignition' shows that the determination has serious limitations. In most cases, when analysing traditional materials, the result obtained from the standard procedure for the determination is meaningful. Within the limits of the data required for control of ceramic production of processes, the small errors introduced by the presence of typical amounts of fluorine and sulphur are acceptable. In specific circumstances, even within this range of materials the loss on ignition figure can be misleading. With more complex materials the loss on ignition figures can be totally confusing and lead to incorrect conclusions. It is sufficient to cite only two somewhat similar instances with materials in regular use in the manufacture of frits, glazes and glasses as a source of borate, namely, colemanite and boric acid. Colemanite is a calcium borate which, on heating, decomposes, releasing not just water but also boric oxide. Boric acid, similarly, releases water on heating, but as it is itself steam volatile, the steam it releases may volatilize some of the remaining boric oxide. Thus, in both cases, the loss on ignition, if combined with determination of boric oxide on the residue after ignition, would give satisfactory analytical totals. However, the boric acid figure would be lower than that available in the material if it were to be mixed with other ingredients and heated to form a frit or glaze as in the production process. If, on the other hand, the correct boric oxide contents were obtained by determinations on the unignited raw materials, analytical totals would be too high by the amounts of boric oxide driven off during the determination of loss on ignition. In any event, the ceramic technologist will not receive the figure for the boric oxide content that is required unless this has been derived from a determination on the sample as received.

'Loss on ignition' is a useful, analytically convenient determination, but the result should be treated with circumspection. Rigid adherence to a standard method of determination without careful consideration of the chemistry of the material can result in disastrous misinterpretations. Additionally, attempts to find modified ignition temperatures and times to allow total completion of one reaction and total non-occurrence of another are usually doomed to failure. Even when handling materials of the same type, a slight change in the composition can result in a change in the initiation temperature of a reaction. Therefore, where problems may arise, it has to be accepted that a simple determination of loss on ignition will not provide an analytical total and its use may have to be abandoned. The accuracy of the remaining determinations may have to be checked in other ways. In one of the instances cited above, viz, colemanite, it is virtually impossible to obtain an analytical total; certainly determinations of water and possibly carbon dioxide would be needed, together with any other potentially volatile impurities that could be detected. It follows, therefore, that in the wider sphere of mineral and ceramic analysis the loss on ignition determination must be used with discretion, and care must be taken to ensure that it is not used outside its valid range.

Having discussed some of the more general problems in the determination of loss on ignition, it is now necessary to look at some of the individual volatile components and their behaviour in a range of the more common materials. This information is needed to enable correct judgements to be made of the validity of the determination for the analysis of specific types of material. The information given below should be regarded as illustration rather than exhaustive. Every analyst needs to note limitations and problems, found as a result of his own experience or described in the more reliable literature.

This chapter is therefore concluded with data concerning the volatile constituents often encountered in general ceramic and mineral laboratories and their behaviour in some of the more common materials.

4.4 EVOLUTION OF VOLATILES

Many raw materials, partially manufactured products and even finished products contain organic materials. Although these generally oxidize during heating to produce water, carbon dioxide, etc., there are instances where some, at least, of the organic content volatilizes before oxidation. A number of shales, exemplified by the Scottish shales, which are often used as fireclays, contain oil, and this can often be seen evaporating as a plume of smoke during heating. Volatile organic materials can be added (intentionally or inadvertently) during the manufacturing process, e.g. as an aid to working or for lubricating presses. Even in the case of finished products, e.g. fibre glass, resin may be used, and although it is unusual for these to be volatile at low temperature, the changing needs of technology indicate that the analyst must bear this possibility in mind. Care must be taken when such volatiles are present. Too rapid heating can result in the volatiles catching fire and, as the situation is no longer under control, loss of sample can occur by spurting.

4.4.1 Evolution of Water

In many ceramic materials hydrogen may be present in one or more of three possible forms, in two of which it is combined already with oxygen. The first of these two takes the form of water of composition, i.e. either as water of crystallization or as hydroxyl groups forming part of the lattice, and the second exists as adsorbed water. The third form of hydrogen is that combined in the organic matter often associated with many ceramic materials, in particular clays.

The first water to be lost on heating comes generally from adsorbed water and from water of crystallization. The process of drying to 105–110°C will release some water (but not all) from a powdered material and water of crystallization from some compounds. Water is adsorbed on to the surface of almost any powder, the amount being conditional on the nature of the material, the specific surface area, the humidity of the atmosphere and the temperature. It is commonly assumed that such water will readily be lost by simply drying the material to a little above the boiling point of water. This is, of course, not true, and in fact for many materials is not even a satisfactory approximation from the analytical point of view.

This may be well illustrated by the example of alumina. Even fused alumina, after grinding finely enough for analysis, (e.g. $\simeq 125$ nm), can give losses on ignition of the order of 1% on heating to 1025°C for 30 min, even after drying overnight at 110°C. It is extremely unlikely that a material that has previously been subjected to a temperature sufficient to melt it, >1900°C, would retain sufficient material volatile at 1025°C to produce a 1% loss in weight, so that this must have arisen from the adsorption of water etc. as a result of grinding. Similar evidence is available from work on a number of ceramic bodies which, after firing to temperatures in excess of 1025°C and then being allowed to stand at room temperature under normal atmospheric conditions, have been shown to retain water to temperatures of several hundred degrees Celsius. Fired and fused aluminas are among those materials for which it is more sensible to report the analysis on an ignited basis and to quote the 'loss on ignition' separately. This course gives more meaningful figures to the ceramist. It is of little value if a nominal 99.8% purity alumina is reported as containing only 98.8% Al_2O_3 plus a one per cent 'loss'. This could give rise to considerable discussion.

Clays are by way of being a special case, but, as many materials contain small amounts of clay or have some clay-like properties, their particular kind of behaviour is important. A pure kaolin tends to have a fairly sharp division between its 'adsorbed water' and its constitutional water. Drying at 110°C can be taken as removing the former while leaving the latter untouched, but even this should not be taken literally. 'Pure' kaolinite is a rare occurrence; almost all clays met with in general practice contain impurities. Disordered kaolin is very common, as are illitic clays.

Montmorillonite represents the opposite extreme, where it is almost impossible to distinguish between the two types of water. There is no temperature at which it is possible to remove adsorbed water without the danger of removing some 'lattice' water. At almost any temperature at which there is a measurable rate of loss of water, it is possible that some of each type will be lost. Increase in temperature will increase the rate of loss, but at any temperature where water is lost the loss tends to continue with time. Although the situation has not been fully investigated from an analytical point of view, it is clear that it is almost impossible to identify baselines with bentonitic (montmorillonite) or similar clays to give either a guaranteed loss on ignition or an analysis. It is not even possible to carry out the loss on ignition determination and the analysis on an air-dried (room temperature) sample, as the adsorbed water content will vary with

LOSS ON IGNITION

humidity. It is normal practice for routine work, when handling clays, to dry the material for at least 4 hours, but no more than overnight, and to assume that such material is 'dry'.

Some clays are extremely hygroscopic, to the extent that it is almost impossible to weigh out a sample. Even using modern electronic or automatic weight-loading mechanical balances it is possible to follow the weight increase as the sample rests on the pan. In this case it may be desirable, and in the extreme imperative, to use a 'catch' weight by tipping out a portion of the sample into the crucible or dish from a stoppered weighing bottle, weighing the bottle before and after.

In these cases weighing is very difficult and can cause problems, particularly where the 'base' state of the sample is indefinite. Then the analyst needs to know exactly what data is required. Often, it may prove necessary to explain the difficulties to the user of the analysis, to make enquiries as to what the data is needed for, and to agree just what figures are to be obtained and quoted.

At the other end of the temperature range, it is natural to assume that ignition to 1025°C or thereabouts for 30 min or longer will remove all the 'water', in whatever form it may be. Again, this assumption must be questioned, absurd though this may appear. Experience has shown that, given the right circumstances and material, the assumption is not necessarily correct. Two examples, both of which have been experienced, may be quoted. First, dealing again with alumina; it is quite possible that some water is retained after ignition to about 1000°C. Reverting to the use of the classical method, when the ammonia group precipitate containing the hydroxides of iron, titanium and aluminium, but chiefly aluminium, was ignited to 1000°C, analytical totals tended to be high, often above the generally accepted maximum of 100.5%. With the advent of 1200°C electric furnaces, a precipitate containing about 40% of Al_2O_3 could lose about the equivalent of 0.5% on heating for 15 min at the higher temperature.

The second example is more obviously chemical in nature. During the analysis of bone ash by either chemical or XRF methods, it has always proved difficult to reach totals of 100%. Bone ash is a difficult material to analyse by 'wet' chemical procedures, the methods being generally poor. However, even when XRF is used the problem remains. Attempts were made to solve it by adding the determinations of carbon dioxide, sulphur trioxide and even chloride and fluoride. The first two have become more or less standard requirements, as these compounds are often present in significant amounts. These figures were added into the total on the grounds that, as the material was basic (viz, $CaO \cdot Ca_3PO_4$), these constituents might not be lost at 1025°C, as the material was a type of apatite. Many forms of apatite could be present, including hydroxy-, carbonato-, sulphato-, fluoro-, etc., which are, of course, more difficult to decompose than the equivalent calcium compounds, so that it is probable that the low totals are due to a failure to break the samples down during ignition. Frequently, and particularly with recently calcined bone, most of the apatite would be in the hydroxy-form. Thus, the −OH would not be identified either as a constituent in its own right or in the loss on ignition. Its presence may, of course, be confirmed by infrared absorption spectrometry. In the case of bone ash, the analyst needs first to appreciate the problem and then to discover whether the ceramist would prefer an actual determination of water, total or residual, after ignition to, say, 1025°C, or simply to have a loss on ignition carried out at a sufficiently high temperature (e.g. 1200°C) to ensure the complete breakdown of hydroxy-apatite.

Finally, organically bound hydrogen will be lost at temperatures well below 1025°C, provided that an adequate supply of air is maintained to oxidize the organic material. This point will be touched on again during the discussion of the evolution of carbon dioxide, since the importance of lack of adequate air relates more to the carbon than to the hydrogen in the organic material.

The material after heating to determine the loss on ignition has to be cooled before weighing. Some of the products may adsorb moisture rapidly from the air. Cooling is usually carried out in a desiccator containing silica gel. Silica gel is a convenient but not very effective desiccant. For most ordinary materials it is quite satisfactory, but again the analyst must consider the material being analysed. More effective desiccants such as phosphorus pentoxide, calcium chloride or magnesium perchlorate may be needed.

4.4.2 Evolution of Carbon Dioxide

Carbon is present in ceramic materials, usually in one of two forms. It may be present as carbonate, most frequently as calcium or magnesium minerals, although siderite (ferrous carbonate) is fairly common in some clays. Additionally it may be present as organic carbon and, infrequently, as carbides. The presence of carbides causes specific problems which are dealt with later.

Carbonaceous or organic matter is found mainly in clays, although it is obvious that any material dug from the earth is likely to suffer from some organic contamination. As far as the loss on ignition determination is concerned, the amounts present offer no difficulty with most materials. It is particularly with clays that care has to be taken, since some clays, can contain significant amounts of carbonaceous material, the so-called 'humic acid'. If the rate of heating is too great and/or there is a lack of readily available air, all the carbon may fail to burn out at reasonably low temperatures.

It is important, however, to remove all the carbon from the clay below about 700°C; otherwise there is a grave danger that oxygen may be taken from some of the iron oxides present, reducing them to the metallic state. The resultant iron may then alloy with the crucible or dish. This results in a high loss on ignition figure, owing to the loss of oxygen, and in the retention of some of the iron in the alloy of the dish even after the fusion, giving a low iron figure in the analysis. There is the added danger that, if this residual alloyed iron is not detected, it will be carried forward to the next analysis. The simplest way to detect it is by heating the dish for several minutes to red heat, which tends to re-oxidize the iron resulting in a characteristic blue–purple stain. Its removal is best accomplished by successive heatings to oxidize the iron followed by dissolution of the oxide with hydrochloric acid $(1+1)$. The danger of the reduction occurring can be greatly decreased by carrying out the whole of the loss on ignition determination in an electric furnace, using a slowly increasing temperature schedule and so yielding a very positive oxidizing atmosphere, e.g. in a tunnel kiln as described in Chapter 2.

In laboratories that are called upon to carry out analyses of materials of low iron content (as well as those of higher iron contents), it is to be strongly recommended that separate dishes be retained solely for the low iron content samples. In fact, this precaution is to be set down in the amendment to BS 1902: Part 9.2, which allows the extension of the method for silica refractories to glass-making sands. The combination of circumstances, where there is a possible high organic content combined with medium or high iron content, requires maximum care. The presence of some of the iron in the already partially reduced state, i.e. ferrous iron, could well increase the difficulties.

China clays are non-sedimentary clays, and have in any event been subject to preliminary mineral separations. The commercial materials tend to be low in both organic matter and iron. Ball clays, on the other hand, are sedimentary in nature and tend to contain 1–2% of Fe_2O_3 together with varying amounts of organic matter (up to several per cent). Some care is therefore necessary in igniting them to ensure the removal of carbonaceous material at a low temperature.

Fireclays are not met with so frequently now as previously: production of firebricks made directly from clay is of much less consequence in the refractories industries as the demands made on the refractories have risen. Also, the production of sanitary fireclay has declined enormously over the last 30 years, vitreous china having mostly replaced it. Nevertheless, fireclays are still being analysed with reasonable frequency and, as one would expect from their association with coal seams, they can be very variable in both iron and organic matter content. The material can contain pyrite, FeS_2, a mineral that seems particularly prone to reduction during the loss on ignition determination unless particular care is taken in burning out the organic material.

Finally, brick-making clays (marls, etc.) can be of various types; they are probably the most variable in composition of all the types of clay used commercially in the ceramics industries. They often contain high iron contents, hence their red burnt colour, and the organic content can vary from very low levels up to significant amounts (as in the Oxford clays). The organic content can be so high that in the manufacture of fletton bricks it provides an important part of the whole energy requirement to burn the brick. Thus, again, in the analysis of brick clays it can be very necessary to take care.

Carbonates generally offer few problems. With high levels of content the temperature must be raised sufficiently slowly to prevent decomposition taking place so rapidly that some of the sample is ejected physically by the evolution of gas. Siderite, a ferrous carbonate often found in brick clays, starts to decompose at low temperatures, and most of the carbon dioxide from this source is lost before the carbonaceous matter has burnt out. Other carbonates, however, do not normally lose carbon dioxide until the carbonaceous matter has gone, provided again that the rate of temperature rise is controlled and adequate air is supplied. Magnesium and calcium carbonates will be completely decomposed during the conventional loss on ignition determination. Even barium carbonate, whose decomposition temperature is higher than the normal loss on ignition temperature, may well be at least partially decomposed if siliceous impurities are present.

Materials high in carbonates pose an additional problem; calcium carbonate decomposes to the oxide, which may then absorb either water or carbon dioxide from the atmosphere, the former very readily. Thus it is imperative to use a good desiccant while cooling the sample prior to weighing. Care may even have to be taken to exclude carbon dioxide from the desiccator if results of the highest accuracy are required. Similar arguments can apply to other basic materials. Bone ash, for example, contains both carbonates and hydroxides, and if freshly calcined the latter can predominate. The ignited material can be very hygroscopic due to possible reaction whereby calcium hydroxide (from hydroxy-apatite) reacts with the sodium carbonate present to produce sodium hydroxide; similarly, the oxide (oxy-apatite) together with moisture can produce the same effect. The product is, of course, highly deliquescent.

4.4.3 Evolution of Sulphur Gases

Very few materials, particularly those used in the ceramic industries, contain elemental sulphur. However, it is not uncommon for sulphur-bearing gases to be lost from the sample at quite low temperatures. Already mentioned above is the loss of volatile organic constituents, possibly containing combined sulphur. Other organic sulphur compounds, although not volatile, can be oxidized at relatively low temperatures, with the formation and loss of sulphur dioxide. Another source of sulphur dioxide at quite low temperatures is pyrite. Other materials often encountered by the laboratory analysing minerals may contain heavy metal and other sulphides, most of which will oxidize at low temperatures. This, in fact, causes serious difficulties in the traditional determination of sulphur, demanding the use of a special technique and a specific reagent mixture (Eshcka's mixture) to carry out the fusion.

As the temperature is increased, the oxidized sulphur salts present in the sample will start to decompose. No consideration need generally be given to the decomposition of sulphites, as these salts occur so rarely that they can be ignored. The only occasion where they may be expected is if they have been deliberately added to the product for specific technical reasons, e.g. as sulphite lye used in the manufacture of silica bricks.

Sulphates decompose at rather higher temperatures than carbonates, but again, reference to the decomposition temperatures of the pure compounds in the literature has little relevance to the actual temperature at which evolution of sulphur oxides may be expected. In practice, when dealing with either fairly siliceous materials, where the sulphates are merely minor contents, or when handling even calcium sulphate containing several per cent of siliceous material, decomposition temperatures will be much lower than theoretical. The most commonly occurring sulphates are those of magnesium and calcium, but occasionally those of the alkali metals or barium and strontium may be present in small amounts. From the above, it is clearly not possible to be dogmatic as to which, or how far each, will decompose at a temperature of 1025°C. Much will depend on the matrix; the more acidic in nature the material, the greater the chance of decomposition. With normal aluminosilicate materials such as clays, fluxes or fired products, one would expect magnesium sulphate to be completely decomposed, calcium sulphate to be partially or even completely decomposed and alkali, barium and strontium sulphates to be slightly broken down. It is possible with the more unusual type of sample that there may be present sulphates of aluminium, iron and the heavy metals; these will be fully decomposed.

It is dangerous to assume that, just because a finished product has been ignited during manufacture to temperatures nominally above the decomposition temperatures of all the sulphates expected to be present, the amount remaining is negligible. In particular, it may be hard to believe that a further ignition of the ground material at a much lower temperature (1025°C) can release further sulphur trioxide. Two factors need to be taken into consideration. First, during the firing of the product, the ware itself may be of large dimensions, for example bricks or sanitary ware, or it may be fired in bungs, as are tiles and plates. In either case, escape of all sulphur gases may be hindered, especially if there are significant amounts of basic materials present. There may, for example, be several per cent of lime present in tiles. The centre of the bung of tiles attains its firing temperature, and either this may be lower than that attained by the outside or, at least, the full temperature may be sustained for a shorter time. Any sulphur released then from the centre has to find its way out of the bung through a matrix containing, say, calcium oxide, only some of which may have reacted to form silicate and be incorporated into the glass of the body. Thus, successive combinations with the remaining calcium oxide and subsequent decompositions will take place, rather like adsorption and desorption of ions as in chromatography. The sulphur trioxide is therefore delayed on its way out into the kiln atmosphere, and as the kiln car is continuing to move towards the cooling zone of the kiln during this time the tile itself is cooling and decomposition becomes increasingly slow. Thus, it is possible that significant amounts of sulphate may remain.

The second consideration concerns the kiln atmosphere and, although it applies to all materials, tends to be of most consequence in the firing of bricks, mainly because many of the clays used contain higher levels of sulphur than the white-burning clays. Two other factors have a bearing, first the dense setting and, thus, the weight of ware being fired at any one time and, secondly, the fact that high sulphur fuels, originally coal and low-quality fuel oils, are more often used to produce the relatively low priced product. The large increase in the price of oil fuels in the relatively recent past has markedly reduced this second factor, firing being increasingly done with such fuels as LPG. However, there are the first signs of a revival of interest in the possibility of firing with coal, again for price considerations. If this happens the problem will almost certainly recur.

High levels of sulphur in the kiln atmosphere, whether from the raw materials or the fuel, suppress the elimination of sulphur from the product, particularly where significant amounts of calcium are present in the body. In fact, a high level of sulphur in the kiln atmosphere, especially when associated with poor ventilation, can even introduce sulphates into a body whose raw materials were virtually free from sulphur before firing.

4.4.4 Evolution of Halogens

The most common halogen is fluorine and it is, in any event, this that gives rise to most ceramic problems and also to difficulties with the determination of loss on ignition. Now that more samples are being received from outside the UK, higher levels of chlorine are often found, particularly in samples from the Middle East. Of recent years the presence of bromine has been discovered in samples associated with the sanitary ware industry. This is because of the growing practice of using polystyrene in the setting of the kiln cars. This polystyrene contains a small amount of a bromine compound used in its formulation. This is released into the kiln atmosphere and thence to the ware and any kiln deposits.

The problems with the halogen compounds differ from those discussed in the sections above in that such compounds do not themselves usually decompose in the range of temperatures normally encountered, but are removed by reaction. Carbonates and sulphates break down, even in the pure state, into acidic and basic oxides at defined temperatures, and most of this will have occurred by about 1200°C. The sulphates of strontium and barium might be considered as exceptions, but only in that slightly higher temperatures would be involved. In the impure matrices normally encountered, decomposition temperatures are often reduced to within the limit of that used for the determination of the loss on ignition. Halogen salts, on the other hand, may begin to volatilize as the

LOSS ON IGNITION 63

actual compound. Release of the halogen or its hydride, i.e. the gaseous acid, normally occurs as a result of reaction with a stable, non-volatile oxide such as silica, usually in the presence of water. There may well be some indication of a reaction by which the hydroxyl group from steam exchanges with the halogen, ultimately resulting in the formation of the basic oxide and the halogen hydro-acid. Thus, the release of halogens or relatively volatile compounds is usually assisted by the presence of other suitable constituents.

Temperatures of such reactions are not easy to predict. They are very dependent on the physical state of the compounds present and the state of subdivision; in other words the juxtaposition of the two or more reacting species. Again, the matrix itself is of significance, more so than in the situations discussed above. Although this applies to each of the halogens, most attention nevertheless needs to be focused on fluorine.

Fluorine finds its way into most pottery body formulations via the use of Cornish stone and china clay. In addition, most other clays contain amounts up to about 0.03–0.04%; slightly lower but identifiable amounts are found in most materials. Experience has shown that process firing to temperatures between 1000°C and 1200°C tends to remove about half the original fluorine. Again, the extent of the loss depends on the bulk of the pieces forming the product, the kiln atmosphere and the ventilation. In the case of fluorine, there is an additional and important factor that comes into play even when compared with the other halogens. A further and dominant reaction can occur with fluorine which does not affect the other halogen compounds. This reaction is the exchange of oxygen and fluorine in a silica molecule to produce volatile silicon tetrafluoride. The degree of loss of fluorine and the temperature at which it will occur are again very dependent on the matrix, more so than in the previous cases. Pure calcium fluoride, for example, would not decompose in normal temperature ranges. In order to achieve decomposition the presence of another non-volatile acid radical such as silica is necessary. Silica, of course, combines not only with the fluorine but also with the calcium, and thus aids the decomposition very markedly (see Eqn 4.3).

The presence of steam in the vicinity greatly enhances the reaction by means of the reactions shown below.

$$CaF_2 + H_2O = CaO + 2HF\uparrow \qquad (4.4)$$

$$SiO_2 + 4HF = SiF_4\uparrow + 2H_2O\uparrow \qquad (4.5)$$

The presence of water in the original materials and the temperature at which it is lost affect the overall loss of fluorine. Clearly, if there is hydrogen (steam) present at temperatures where the second pair of reactions (Eqns 4.4 and 4.5) occur, the fact that both products on the right of Eqn 4.5 are volatile will tend to make the reaction go in that direction, provided that adequate ventilation removes them from the vicinity.

In carrying out a loss on ignition determination ventilation is very good, and the layer of material is thin, so that gases can escape very rapidly without necessarily reacting with any of the other constituents present.

The presence of boron in the sample will also tend to cause the removal of fluorine as boron trifluoride, again provided that the matrix is not appreciably alkaline. This particular subject will occur again in the chapter on fusion (Chapter 5), where it will be discussed at greater length.

4.5 CONCLUSION

It will be seen from the foregoing that the determination of loss on ignition may be a simple procedure, but it can entail a wide variety of complex and complicated chemical reactions. Most materials in common use for the manufacture of ceramics have relatively simple decomposition patterns and relatively low levels of potentially volatile acidic impurities. For these materials, a determination by a standard procedure, can, and usually should, produce figures that can be related both theoretically and practically to an analytical total. This means that all the decomposable entities have effectively been decomposed and potential reactions producing losses in weight have gone to virtual completion, within the limits required by the analyst. The residue from the ignition process, on which the analysis is often conducted, consists essentially only of oxides normally determined in the class of material in question. In these cases, failure to achieve a satisfactory analytical total can usually be attributed either to a failure to identify and determine an additional and unexpected oxide or to analytical error.

It has to be admitted that the above statements are not absolutely correct. There will remain in some materials some sulphur as sulphate, fluorine as fluoride and even, as in some apatites, carbon dioxide or hydroxide. In some cases an increase in the temperature of ignition can often eliminate these constituents, but at the possible risk of losing some of the alkali content. For most common materials, however, the level of content of residual volatiles is so low that it is best ignored.

As the complexity of the material increases, so the value of the loss on ignition determination decreases. Once significant amounts of sulphate or halide, particularly fluoride, are present it becomes virtually impossible to devise an ignition procedure that will produce a meaningful residue, free from potential 'volatiles'. Analytical totals can be achieved only by analysing the actual residue for its volatiles content. This is extremely time consuming and would in any event be unprofitable. If the analyst is forced to the lengths of determining the content of the various potentially volatile elements, it is surely more relevant to carry out the determinations on the original sample, where the results might conceivably be of some value. With complex samples, therefore, it often has to be accepted that an analytical total cannot be obtained for use as a cross-check on the correctness of the other determinations. Alternative approaches have to be made to ensure the reliability of the analysis. Where the type of sample is regularly analysed and the loss on ignition cross-check can be used, the simplicity of XRF procedures is such that duplicates are normally not necessary except for referee work. Errors, if made, are usually gross and therefore self-evident. Complex samples, for which the loss on ignition approach is not possible, will normally require the performance of at least duplicate determinations.

5 Decomposition of Samples by Fusion

5.1 INTRODUCTION

Why fusion? The process of decomposing samples by fusing them in a flux is a relatively time consuming procedure, in addition to using expensive platinum apparatus and costly pure reagents. These disadvantages have been set against and presumably outweighed by corresponding advantages, otherwise fusion techniques would not have proved so universally popular. If reproducible and accurate analysis of the types of samples dealt with in this book is to be carried out, certain essential conditions have to be met with in regard to the presentation of samples to the instrument. It is because these conditions are met by fusion techniques that fusion is so universally employed.

(1) Samples have to be presented in a homogeneous form. Unfused samples of identical composition may still differ from each other in both mineralogical make-up and particle size. Either of these differences can produce differences in elemental count-rates. Particle size differences can be minimized by instituting rigid control of grinding procedures but, even though maximum care is taken, variation cannot be completely eliminated. Very fine grinding is, in any event, essential and this results in other problems. Heat produced during the grinding process can result in the loss of volatile material. The main problem, however, is that the level of contamination from the mortar greatly increases as the particle size is reduced. Although no precise figures are available, it is reasonable to hypothesize that the amount of contamination will increase with the surface area. Thus, the error introduced will not be directly proportional to the degree of grinding but will increase by about the inverse square of the particle size expressed as a nominal diameter. In the case of the McCrone grinding mill used for preparing very finely ground samples for X-ray diffraction (XRD) analysis, significant contamination from the grinding elements has been observed.[1]

Calibration problems, using unfused samples, mean that accurate results will be achieved only if the mineralogical composition of the samples is virtually identical with that of the materials used for the calibration. Thus, calibration using ground, unfused samples can only be used with any accuracy where both standards and samples are from the same mineralogical source. Even so, caution is necessary in interpreting results obtained using such calibrations, as there is always the possibility that, even in the same quarry or seam, differences in mineralogy may occur. In any event, separate calibrations are needed for each source of sample; this could add up to very large numbers of calibrations having to be maintained.

Fusion affords consistency both in particle size and mineralogy. The effective particle size after fusion is approaching atomic, provided that the degree of dilution is adequate. This also means that the mineralogy of the sample is totally destroyed, so that the analyst is dealing with the same basic matrix at all times. The sample is presented to the instrument in the form of a glass of reasonably constant composition, effectively a solidified solution. The differences in composition are of minor components of a lithium borate glass, i.e. the components to be analysed for and related to the original sample.

(2) Dilution of the sample also has the effect of reducing inter-element effects and, by selecting an adequate but constant dilution, the magnitude of these effects may be readily established and corrected for. If dilution is carried out by the use of very light elements, the loss in sensitivity is minimal and interference from the elements in the flux can for the most part be ignored. Hence the great value of lithium borates as fluxes.

(3) Presenting fused samples to the instrument allows standards to be prepared from pure oxides, carbonates, etc. so that calibrations can be prepared synthetically for almost any type of sample. Standards can also be prepared to establish the exact values of line overlaps and of inter-element effects. This avoids the need for the use of analysed reference materials, with their inherent possibilities of slight inaccuracy. From this it follows that there is no need to apply multiple regression analysis in order to establish calibration curves. This latter practice is necessary when materials such as metals are to be analysed, as synthetic standards cannot generally be used. In fact, the use of multi-regression analysis has been the norm, and it was only widely appreciated comparatively recently that it is not necessary with materials that are analysed by a fused, cast bead treatment.

The advantages of direct calibration by preparing standards at uniform intervals throughout the range of interest are significant. The curves are by definition more accurate, given careful preparation of the beads, than curves prepared from beads using analysed standards, even if they are Certified Reference Materials. Also, any synthetic bead happening to contain an error (e.g. in composition) shows up immediately by its deviation from the calibration graph. Any such point can then either be ignored, or better, a new standard bead prepared and the point checked. In the unlikely but possible event that some odd and unexpected interference is occurring, the deviation will still be present and will demand detailed investigation. Normally, however, the error will disappear. Using multiple regression techniques and a computer, the latter

DECOMPOSITION OF SAMPLES BY FUSION 69

will accept all the data and the calibration will incorporate an erroneously made up bead. Only analysts who are prepared to draw graphs manually, showing the actual points, are likely to notice the problem. It should be noted that some instrument manufacturers may still attempt to sell a standard package, including multiple regression calibration. This, to them, may be the norm, but it is wise to insist in having the more direct and better system as an additional option. Fortunately, there is an increasing tendency for instrument manufacturers to supply software which permits plotting of raw calibration data.

The ability provided by fused, cast beads to use synthetic calibration has further and profound consequences that are not immediately apparent. The use of synthetic standards made from pure materials makes valid the claim of XRF to be a primary method and suitable for consideration for standard work. XRF is thus placed in exactly the same position as other accepted techniques, such as flame photometry, atomic-absorption spectrophotometry and spectrophotometry, all of which depend on the preparation of standards from pure reagents for their calibration.

This acceptance of XRF as a standard method is important to industry for several reasons. As time passes fewer and fewer analyses are being carried out in this field by techniques other than XRF. This, at least, would make it very difficult to prepare Certified Reference Materials in the future. These can only be established by using the results obtained by several reliable laboratories, using techniques of acceptable accuracy. If there is an insufficient number of laboratories using such acceptable techniques, i.e. standard methods, there will be no more CRMs. The acceptance of an XRF technique avoids this problem.

The XRF method is simple and, once a laboratory has acquired the right skills, reliable results can be readily produced. As the laboratories of the UK ceramic industries adopt XRF, they also tend to adopt the co-operatively developed standard methods, the first of which are now available (BS 1902, Parts 9.1 and 9.2). If they are using these methods on a routine basis, it may be claimed that the products are subject to analytical quality control using BS methods. This, in turn, has considerable commercial advantages. Finally, laboratories providing a service to industry and accredited under such schemes as NAMAS, and using XRF procedures, are able to offer the use of BS methods for their analyses. Not only is this important for the general run of clients, but it is imperative to offer such a service when clients have agreed in advance with a third party that standard methods of analysis shall be used. In the latter circumstances, reversion to 'wet' chemical methods makes the cost enormous and usually prohibitive.

5.2 CHOICE OF FLUX

Early fusion methods concentrated on using borax as the flux. At that time, sensitivity for sodium was so poor that it was not determinable by XRF, so that using a flux containing the element imposed no penalty. In addition, borax was much cheaper than lithium fluxes and more readily available in a pure form. However, borax is not ideal. It suffers from the considerable disadvantage that beads made from it cannot be stored for any length of time. They tend to deteriorate by picking up moisture from the atmosphere, so that standards have to be replaced at too frequent intervals.

The development of acid phthalate analysing crystals, in particular that of thallium, meant that sensitivities for sodium increased to the point where its determination moved from pointless through feasible to eminently practical. This, of course, sounded the death-knell for the use of borax as a flux for general analysis of rocks, minerals and ceramics. Some laboratories with special requirements, not including the determination of sodium, still use borax, counterbalancing the problems of bead stability against the considerable savings in cost of flux; lithium fluxes are still very much more expensive for similar levels of purity.

As these changes were taking place there was, in any event, increasing interest in the use of lithium borate fluxes. This was encouraged by the widespread use of, firstly, lithium tetraborate and then metaborate for AAS analysis of geological and ceramic materials. However, beads made from lithium metaborate, although more stable than borax, still showed a generally unacceptable lack of stability to atmosphere. Hence, most analysts turned to the use of lithium tetraborate, which produced beads of significantly greater stability.

As development work at Ceram Research progressed, it was realized that tetraborate had two important disadvantages as a general flux. Its melting point is so high that the melt of a sample in tetraborate only becomes really fluid at 1200°C in a furnace. When used on gas burners the melt remains very viscous; too viscous, in fact, to swirl readily so as to help distribute the sample evenly throughout the melt. As swirling to obtain good mixing is an essential part of the fusion process, this meant that it could only be carried out whilst heating in a 1200°C furnace and repeatedly removing the dish in order to swirl. This was not only time consuming, it was also wearing, not to say positively painful, for the analyst, who was forced to work in front of an open furnace at regular intervals during each fusion. A second problem that arose as a result of the use of lithium tetraborate as a flux was the fact that highly siliceous materials were very difficult to decompose. It was found that an addition of some metaborate to the flux made the decomposition of such materials much easier and faster.

5.2.1 Theory of the Suitability of Fluxes

As a result of noting the above facts, experiments were carried out in which three fluxes of interest, lithium metaborate, lithium tetraborate and a mixture of the two, were used to decompose a wide range of samples. Silica/alumina range samples from high silica to high alumina content were at the time the main interest, as they constituted the bulk of samples. It was found that lithium tetraborate, as already stated, only decomposed high-silica materials with difficulty, whereas metaborate took them into solution with ease. Lithium metaborate, on the other hand, was ineffective for high-alumina materials whereas tetraborate was eminently suitable.

The mixture of lithium tetraborate/metaborate chosen was almost the eutectic mixture ratio, which is almost exactly four parts of metaborate to one of tetraborate by weight. This mix proved an ideal compromise, bridging the gaps left by each individual flux. It decomposed the whole range of silica/alumina materials with little difficulty; in fact, the use of the 1200°C furnace was only essential when the sample contained corundum. Most materials seemed to be completely decomposed after a few minutes on the gas burner with occasional swirling. However, if the method is to be adopted as a routine procedure and all silica/alumina materials are to be analysed using the same calibrations, then it follows that a period of treatment in a 1200°C furnace should be used for all samples to make consistent any volatilization of the flux constituents. Failure to give the furnace treatment to samples containing corundum may result in devitrification of the beads with particles of unfused corundum acting as nuclei.

The mixed flux also has the same advantage as metaborate alone, it is fluid at the lower temperature produced by the gas flame. In fact, as might be expected for a near eutectic mixture, the mixed melt is even more fluid than metaborate alone.

Extending this work to cover carbonate materials such as limestone, dolomite and magnesite showed that the best flux, without doubt, is lithium tetraborate, even allowing for its high melting point. This is helped by the fact that the lime in the samples serves to lower the melting point and increase the fluidity of the mix by the formation of calcium borates.

Looking at these results it became clear that some general conclusions were drawn concerning the correct flux to use for each type of material.

There is a long-standing ceramic convention in the formulation of glazes concerning acidic and basic oxides. Silica and boric oxide, for example, are regarded as acidic oxides and the metallic oxides are considered basic. Basic and acidic refer here, of course, to their high-temperature behaviour, where because of their involatility they are 'stronger' acids than, for example, sulphuric acid; this reverses the normal room temperature situation. This division into acidic and basic oxides seemed to be relevant to decisions concerning choice of a correct flux.

It is possible to think of the two primary fluxes under discussion as combinations of the oxides of lithium and boron, as is conventional in ceramic nomenclature generally. Thus:

Lithium metaborate = $Li_2O \cdot B_2O_3$
Lithium tetraborate = $Li_2O \cdot 2B_2O_3$

It is possible, therefore, to think of metaborate as a relatively alkaline flux, as it contains one alkaline oxide to one acidic oxide molecule. Tetraborate is relatively acid, as it contains two acid oxide molecules to one basic. Reaction is most likely to occur between acid oxides and a basic flux and, conversely, between basic oxides and an acid flux. Silica has to be considered as a strong acid oxide at high temperatures, whereas the alkalis and alkaline earths must be considered strongly basic. Alumina is normally considered as an amphoteric oxide and will react, in general, as the opposite of the matrix in which it finds itself. Experience has shown that, in terms of fusibility in the fluxes, alumina tends to react as rather more basic than acidic. The mixed flux will, if anything, be expected to have a tendency towards basicity.

The arguments above suggest that selection of the flux should be based on a consideration of the acidity or basicity of the material to be fused in the context of this concept. In practice this has proved eminently successful. Consideration must always be given to the possibility of using the eutectic flux, which has proved surprisingly versatile and appears to be able to cope with most materials except for very basic types. Even slags, which are generally high in calcium oxide content, and calcium salts such as gypsum or fluorspar respond to the mixed flux. This flux, as a result, has found great popularity with analysts in the UK, and has been made available by Johnson–Matthey under the name Spectroflux 100B. For convenience, this flux will usually be referred to as 4/1 flux.

If the fused, cast bead method is to be used within a laboratory that analyses a wide range of different types of samples, it will soon be appreciated that it is highly desirable to keep the number of fluxes used to a minimum. As many types of sample as is sensibly possible need to be handled by the most common flux. If almost all samples can be handled by using, on the one hand, 4/1 or, on the other, tetraborate alone, the analyst has a relatively simple problem in ensuring smooth running of the laboratory. In practice this appears true, as only a relatively small percentage of routine samples received in the authors' laboratory has to be treated as 'specials' from the point of view of the choice of flux.

A table showing appropriate fluxes for a wide range of materials is given later in the book (Appendix II). Fuller details for specific materials are given in Chapters 12–19, where full working details enabling the analyst to select all parameters suitable, including flux, are given.

5.3 DILUTION OF SAMPLE IN FLUX

Different sample types not only demand different fluxes, they may also demand different dilutions of sample to flux. Most silicate or carbonate materials can

DECOMPOSITION OF SAMPLES BY FUSION

be handled by dilution with five times their weight of flux. The most common exceptions to this are materials containing very high levels of magnesium, such as magnesites. As was discussed in the previous section, it is highly desirable to keep the number of variants of the basic technique to a minimum. The choice of degree of dilution is thus important, as it must form a compromise between that which gives the highest sensitivity, albeit with acceptable interferences, and ease of decomposition for as wide a range of sample as possible.

Before discussing this aspect in more detail, it is desirable to pause to consider what exactly the analyst is aiming to do. At the risk of being accused of stating the obvious, it is necessary to define what exactly is meant by the *degree of dilution*, as experience has shown that not all analysts appreciate precisely what they are setting out to achieve and therefore may miss some of the fundamental requirements.

In presenting the fused, cast bead to the instrument, the analyst is not, in fact, analysing the sample itself. It is the glass of the bead that is being analysed for the oxides sought. The fundamental assumption is that the final (i.e. ignited) weight of sample is mixed with the final (i.e. dried) weight of flux in exactly the same proportions as in the standards. In order to produce accurate standards, it is necessary to ensure that the final bead contains the correct weight of ignited oxide(s) in the correct weight of flux. Both the oxide(s) and the flux require 'drying' or some form of pre-treatment to ensure that the finally bead contains the correct ratio of flux to oxide. This is also equally true of samples.

If the analyst is aiming to produce a bead wherein the sample is diluted in the ratio of 5:1 with flux, then for the method as normally described in this book, for each 1.5 g of sample, 7.5 g of flux is required. Taking the case of the flux first, when the 1 kg tin of flux is opened, it will probably have stood for a considerable time since its delivery from the suppliers. Although it will be nominally sealed with tape it would be rash to assume that the seal is airtight. In any event, after the material was melted by the manufacturer it has had to be ground and packed, and as a result it has been in contact with the air for some time, partly in the finely ground state. One can expect, therefore, that it will have picked up some moisture from the air, and experience shows this to be true.

Thus, if 7.5 g of flux from the bulk is weighed out, the actual weight of flux after fusion (ignoring any possible volatilization) will be less than this figure. In order to obtain the correct weight of flux in the melt, either of two courses can be adopted. First, an appropriate weight of the flux can be ignited to 700°C in a large platinum dish to remove any picked-up moisture. This material can then be used immediately after cooling, or it can be stored in air-tight conditions in a suitable container for limited periods of time and used as the 'dried' flux. Alternatively, a *representative* portion of the flux in the tin can be ignited to 700°C and the loss of weight noted. A corrected weight of flux is then used with each sample to allow for the loss in weight on heating. It is highly desirable to mix the contents of the tin thoroughly* to ensure even distribution of the moisture before determining the loss in weight, so that the figure obtained is truly representative of the whole amount involved. Unless the contents of the tin are used within a few days it is necessary to check this moisture figure. Weekly intervals will probably be sufficient, but it is best to check that this interval is adequate within the actual environmental conditions of the laboratory.

Having ensured an exact and constant weight of flux, it is now necessary to ensure that the same applies to the second ingredient of the melt, viz, the sample. If, as is usual, the sample loses weight when it is heated, then the total weight of the melt will not be the sum of the original ingredients even when the change in weight of the flux has been allowed for. If one takes 7.5 g of flux (ignited) and 1.5 g of sample to achieve a 5:1 dilution in the melt and the sample itself loses weight during the ignition process, then the final dilution factor will be incorrect. Consider only, for the moment, the losses due solely to heating to the loss on ignition temperature, and ignore any possible further loss resulting from the fusion process. If the loss on ignition figure is 10%, then 1.5 g of sample will have lost 0.15 g in weight during fusion if the unignited sample is used for the fusion. Thus the melt, instead of weighing 9.0 g including 1.5 g of sample, i.e. a dilution ratio of 5:1, will contain 7.5 g of flux and 1.35 g of sample, making a total weight of 8.85 g at a dilution of almost 5.6:1. It is clearly inadmissible to compare this melt with standards made at a dilution of 5:1. Thus, if the calibration is to be strictly applicable, it is imperative either to use ignited samples or to adjust the weight of 'raw' sample introduced into the melt so as to finish with the equivalent of 1.5 g of ignited sample. As will be seen below there are other ways of approaching the problem, but they are time consuming; also they lack elegance and will potentially be marginally less accurate.

Some samples will lose additional volatile material during the fusion. This may happen simply because the temperature of fusion is higher than that at which the sample has been previously ignited. This may arise from further loss of water, carbon dioxide or sulphur trioxide, etc. For example, some of the carbonates, e.g. those of barium and strontium, are unlikely to be decomposed completely at 1025°C in most matrices. However, in lithium tetraborate, or even the mixed flux, they could lose most, if not all, of their CO_2 at 1200°C, because of the replacement of the carbon dioxide by the less volatile boric oxide. Other materials can also react fairly readily with the flux, whereas if heated on their own they might be relatively stable. Examples of this are materials containing fluorides such as enamels, calcium fluoride etc. In these cases, particularly in the absence of major quantities of silica, the fluoride salt may not readily decompose alone, but if mixed with borate, and particularly in a matrix which might be considered acidic, i.e. containing tetraborate, boron

---

***CARE**: Some borates are defined as harmful and threshold limit values are low. When mixing, dust in the air should be avoided.

trifluoride (BF_3) may be lost. In this case, weight is being lost from both the flux and the sample. The most difficult case, and fortunately a rare combination of circumstances, can arise when the slight loss of lithium and borate from the flux can behave like a carrier and assist the removal of other marginally volatile constituents. The phenomenon can, in fact, be observed in the analysis of feldspars and other high-alkali materials. Experimental work has demonstrated that prolonged heating at 1200°C of a melt of this type of material produces a progressive reduction in the alkali contents with increasing time at the high temperature. Fortunately this loss is so slow that the normally used conditions of fusion do not produce any noticeable change in the analytical results.

Fortunately, most routinely analysed types of sample do not give rise to significant problems of this sort. Sample/flux ratios in such cases can effectively be taken for granted, but in more complex materials it may well be advisable to check melt weight after fusion.

5.4 STANDARDS AND SAMPLES

Because the analyst is usually analysing the glass of the fused, cast bead, the analysis of the sample contained in that fused, cast bead can only be obtained by one fundamental assumption. This assumption is that the bead containing the sample has *exactly* the same dilution as all the standard beads containing the oxide mixtures. The standards are, in fact, prepared so that the total weight of oxides contained therein equals, say, 1.5 g (in the case of silica/alumina range materials), thus representing a hypothetical sample. As will be described in the appropriate chapter on standards and calibration, the standard bead will contain the desired amount of element or elements, calculated as oxide(s) to furnish a calibration point, and the remainder of the 1.5 g will usually be made up of one of the main oxides in the material to be calibrated. Silica is used for this in siliceous matrices, except, of course, when calibration for silica is being undertaken.

For the assumption to be true, not only must the weight of standard oxides or sample in the final melt be correct, but also the final *ratio* of standard oxide or sample to flux. In the previous section, many of the factors affecting this final dilution have been dealt with in some detail. The analyst has full control in ensuring that the correct weights of the raw materials of the melt are added to provide the correct ratio in the melted bead. Factors concerning volatilization caused by additional losses in weight by heating the sample to a higher temperature or reaction with the flux are much more difficult to control. The effect of these factors, in those cases where they may be expected to occur at possibly significant levels, can be reduced to an acceptable level by the technique of cooling the melt and weighing it, followed by introducing a melt weight factor into the calculations on the computer to allow for this change in weight.

This melt weight technique is a slow process, as it involves the need to cool the melt prior to weighing, followed by re-heating at 1200°C and then casting the bead. Its disruptive effect on the smooth flow of work is not inconsiderable, diverting attention from other samples. A further possible, but admittedly not tested, disadvantage may exist. There is always the risk during the re-heating process after weighing that further volatilization may occur, which is then not accounted for.

A possible alternative method for correcting for volatilization has been suggested, but not so far investigated. This is to include in the melt, either by a weighed addition or incorporated into a specially prepared flux, a known amount of an appropriate oxide to act as an 'internal standard'. This oxide should preferably be of medium atomic weight and rarely sought in analysis. It should not be so light as to preclude its accurate determination, nor too heavy, so that it causes significant reduction in sensitivities for the elements sought. If possible it should create few interferences via line overlap and relatively small inter-element effects. Then, by measuring the apparent concentration of this element in the bead as part of the analysis, it should be possible to correct for volatilization losses by the deviation between its apparent and real concentration. It is true that this course would add a further error due to the experimental errors in determining the concentration of the element in the bead. However, although this would probably decrease the accuracy of analysis of simple samples, the degree of error is definable. This degree would almost certainly be less than the possible, and not easily identifiable, errors caused by the bead weight system with complex samples. It would also be necessary to take into account the very considerable savings of time to be achieved by the addition method on those samples necessitating the use of the bead weight technique.

There is, of course, a further source of volatilization, namely loss of weight from the flux components themselves. This has been shown to be a relatively small error and has been generally ignored by most analysts. Again there is no simple solution. Theoretically it would be possible to weigh every melt, but even so it is rarely possible, even with apparently straightforward samples, to be totally sure that the loss in weight originates solely from loss of flux.

It is correct practice to attempt to minimize this effect by standardizing conditions during the fusion and casting process as closely as possible. Experiments have demonstrated very clearly that flux loss does not appear to occur at any significant rate while the melt is being heated over gas burners (i.e. at temperatures around 950–1000°C). Losses start to occur when the melt is transferred to the furnace at the higher temperature of 1200°C. As has already been said, this loss is not very rapid and should not affect the validity of the analysis, provided that reasonable care is taken to keep the time of heating within normal experimental control. This should result in very similar levels of volatilization from standards and samples. This control includes the need for having well fitted lids, correctly in place, while the melt is in the furnace. Here, as in so many other instances in XRF analysis, accuracy can be achieved with so little extra effort that poor standards of analysis are inexcusable.

It is possible to visualize, as borates are steam volatile, that the presence of different amounts of adsorbed moisture in various tins of flux could produce different amounts of volatility in the flux. No evidence has ever been obtained to indicate that this effect, if it exists, introduces any identifiable errors into the analysis. The same sort of effect would arise if unignited samples, capable of releasing steam, were introduced into the beads, rather than using previously ignited material. This, bearing in mind that the sample could also lose fluorine, sulphur gases, etc., some of which might react with the flux and cause loss in weight from both volatility and reaction, is a further argument for the technique of using ignited samples in the beads wherever possible.

Some indication of the magnitude of these effects is shown below as derived from experimental work carried out at Ceram Research.[2]

The process of fusion involves subjecting the mixture of sample and flux to high temperatures for varying periods of time. This time can be divided into two parts, first that spent over the burners and, second, that spent in a high temperature furnace, usually at 1200°C. During these treatments two effects can occur to affect the results:

(1) loss of flux; this has the effect of increasing the concentration of the sample in the melt and thereby increasing the recorded analytical results, and

(2) loss of components from the sample; volatile components, most commonly alkalis, can be lost during the fusion process.

There does not appear to be any significant loss of flux or sample during heating over burners within periods ranging from 15–45 min. This gives more than adequate latitude for normal laboratory purposes. Unfortunately the same is not true when heating in a furnace at 1200°C, even with the lid fully in place.

Loss of flux is evident as a result of heating to 1200°C; this loss is actually visible as a smoky vapour arising from the dish when the lid is not fully in place. Periods of time in the furnace used were 5, 15, 30 and 45 min, each after a preliminary treatment of 15 min over burners. After the designated time the melts were removed from the furnace, allowed to cool, weighed, re-heated in the furnace for exactly 5 min and then cast. Although losses of weight are not easily reproducible, the trend of a steady loss of weight was clear. The loss of melt appears to be about 0.002 g per 5 min. The resulting increase in analytical result at the 100% level would be 0.02% per 5 min.

Loss of volatiles from the sample can take place in the 1200°C furnace. Potash appears to volatilize at a rate of 0.006% of K_2O per minute at the 2.64% level in BCS 269 Firebrick. If it is assumed that a proportionate loss will occur with all potash contents, this would result in a loss of K_2O in a feldspar containing, say, 12% of about 0.03% per minute. It is reasonable to assume that similar losses will occur with soda; unfortunately, the errors inherent in its determination precluded their quantification.

Using an RF induction fusion apparatus at a setting said to yield 1300°C reduced the apparent soda content (at about the 10% level) in a soda feldspar by just under 0.1% in 5 min and about 0.15% in 15 min, and potash in the sample was reduced by between 0.76 and 0.69% after 15 min. This suggests that the loss of alkalis could be a problem if the use of an RF induction fusion apparatus is being considered. In the experiments described here, a lid was used on the fusion vessels, which would inhibit the loss of alkalis. Many automatic fusion devices do not use a lid (although the casting mould may be suspended above the fusion vessel), so that significantly greater losses would be expected.

For types of sample containing high alkali contents, control of the time in the high temperature furnace to within say, 2 min is very desirable. Losses of alkalis from the sample should then be adequately matched by similar losses from the standards. Errors due to losses of flux etc. are normally ignorable, contributing only a very minor part of the whole experimental error. Nevertheless, where work of highest precision is called for, correction of melt weights will make slight improvements (but at a high cost in throughput).

5.4.1 Standards

Most standards are prepared from oxides or carbonates, and these materials are normally used in finely divided form so as to facilitate dissolution. Almost all powders adsorb moisture from the atmosphere, and many are strongly retentive of this moisture to high temperatures. In order to ensure that the correct weight is actually incorporated into the bead, most materials used as elementary standards require some form of pre-treatment. In addition, it is not uncommon for those elements that have oxides in different oxidation states to be supplied in one form, whereas the form calibrated and existing in the bead may be different. For example, manganese is most commonly available as the dioxide, but is incorporated in the standard bead nominally as Mn_3O_4. The dioxide decomposes on heating to 1025°C to the latter oxide, and is normally reported as such. Finally, some oxides that are supplied nominally in a specific oxidation state can actually contain a mixture of oxides. If pre-treatment such as heating to a specific temperature can remove the doubt, such as in the case of cobalt, this solves the problem. Where this is not so, recourse may have to be made to an actual determination of the elemental content, although this should be avoided if at all possible. Details of these pre-treatments and discussion of individual problems will be found in the chapter on calibration (Chapter 9, particularly Table 9.1). Carbonates are usually 'dried' at a temperature high enough to ensure that correct weights of dry material can be taken, but not high enough to risk decomposition. The carbonate itself is normally incorporated in the bead mix, decomposition to the oxide taking place during fusion.

5.5 CASTING

It would be possible to enumerate the arguments for or against the presentation of top or bottom surfaces to the X-ray beam. However, as equally good results can be obtained from either surface by experienced laboratories, a theoretical discussion at this point seems to be out of place. The authors believe that the use of the top surface is to be preferred if only for its ease of use. In addition, during the experimental work described in the above reference,[2] facts emerged that suggested difficulties with the bottom surface, and an outline of some of these findings is given below.

If the bottom surface is used and the casting moulds are in good condition, the surface presented to the beam will be flat. On the other hand, the top surface will be convex and will tend to be closer to the beam. Thus, if the top surface is to be used, the volume of the melt (and thus presumably its weight) will determine the thickness of the bead and its convexity. Experimental results have shown that the difference in apparent result between the top surface and the bottom surface using the same calibration amounts to about 0.7%. This serves to indicate an absolute maximum error caused by any deviation in convexity. The effect of intensity with curvature is more pronounced the longer the wavelength (the lighter the element) measured.

If the bottom surface is used, repeated casting in a single new mould gives, as might be expected, better precision than if a series of moulds is used. This is especially true if the series of moulds varies in length of service. For silica, for example, casting in a single mould yields a standard deviation of about 0.05%, as against 0.15% for a variety of dishes.

5.5.1 Effect of the Weight of the Cast Bead

This section is only of consequence if the top surface is chosen for analysis. If the bottom surface is to be used the discussion of bead weight is of no relevance.

Apart from variations caused by loss of weight of the melt by volatilization during fusion, the final bead weight will vary according to how much of the melt is transferred from the fusion dish to the casting mould. There are several reasons why the weight of the bead after casting will vary, even when good analytical practices are being rigidly adhered to. These include the following.

(1) *Melt temperature.* The hotter the melt the more fluid it will be, other things being equal. It is also true that, in general, the amount pourable into the dish will increase with the fluidity of the melt. The temperature of the melt at the time of pouring will depend upon two things:

(a) the temperature imparted to the melt from the furnace. This, normally can be taken as the temperature of the furnace.

(b) the amount of heat lost by the melt between its removal from the furnace and its being poured. This will be primarily affected by the speed with which the analyst completes the pouring. To this extent it is analyst-dependent, and experience shows that differences can be identified in average bead weights between analysts. Training in developing an identical technique between analysts is important. If automatic fusion is used, transfers from fusion dish to casting mould will be virtually identical in time delay and will in any case be carried out with less delay, so that the melt will be more fluid and melt weights more consistent.

(2) *Adhesion of the melt to the fusion vessel.* Platinum/gold dishes, when new, are almost completely repellent to the melt. Progressive loss of gold and roughening of the surface with use tend to cause a greater weight of melt to adhere to the sides of the dish, and this will not be transferred to the casting mould. The melts of different types of sample will tend to adhere more or less to the fusion dish depending on the nature of the material.

(3) *Viscosity of the melt.* The more viscous the melt, the less will normally be transferred. This is primarily a function of the flux and its temperature (see above). A lithium tetraborate melt will be much more viscous than a 4:1 lithium meta/tetraborate flux at identical temperature. The nature of the sample will also, however, cause smaller, but still significant, variations in viscosity. Calcium silicate samples, for example, produce a much more fluid melt than do samples of zircon. In fact, calcium silicate samples yield a mean bead weight of 8.9 g (nominal 9 g), with a standard deviation of 0.05 g, as against similar figures for zircon samples of 8.5 g and 0.34 g.

This variation of the amount transferred to the casting mould means that beads presented to the instrument will vary in convexity of the top surface. With a constant contact angle of ϕ, if the level of the melt lies at any position on the radius of the chamfer between the horizontal portion of the upper flange and the line of the side of the mould, the higher the level in the casting mould, the greater the curvature of the top surface of the bead. This effect will decrease the distance of the nearest point of the bead to the X-ray tube and thereby tend to increase the fluorescent X-ray signal, particularly for the lighter elements. Thus, the greater the weight of the bead, the higher should be the count-rate and, hence, the apparent result. The rate of increase in apparent silica content appears to be about 0.09% per gram of bead weight for levels of content between 50 and 60%. As the spread of bead weights in practice is about 0.5–0.7 g, then errors from this source may be expected to be about 0.05% at the 50% level i.e. almost 0.1% at 100%. No similar conclusions can be drawn for alumina, as results are too imprecise, but the same type of error may be expected, possibly at a lower level.

As bead weight has an effect on the apparent result, the effect on the figures of applying a correction should improve precision. In experiments carried out by fusing BCS 269 Firebrick to produce various bead weights, the standard deviation for silica results (100 s count) across a range of 3 g in melt weight was 0.09%, as against corrected figures of 0.024%. The latter figure compares with a routine s.d. of about 0.06%.

(4) *Deviations in casting procedure.* Newness of dishes, etc., can cause variations in bead weight of about 0.5–0.7 g, leading to a spread of results of about 0.15% at the 100% level. These errors can be corrected for, if necessary, by calculation based on bead weights or using a combined fusion dish/casting mould.

5.6 FUSION OF NON-OXIDE MATERIALS

Modern ceramics include a number of non-oxide materials. Of these, silicon carbide has been in use for a long time, whereas most of the other materials, silicon nitride and its variants, are more recent. So is the need to analyse for the impurities in silicon 'metal' used in the production of a number of these specialized ceramics. These types of material are of growing importance. Silicon carbides, together with graphite, promise to become the main sorts of refractories used by the steel industries. Silicon nitride is currently among the front runners in the field of engineering ceramics, and is finding increasing use in engines. Enormous efforts are being made to extend the use of these materials in engines for both the aircraft and motor industries. Their ability to withstand very high temperatures is a prime factor but by no means the only one.

Silicon carbide materials, in particular, have undergone many changes in formulation during the past few years. In the past, a silicon carbide refractory or abrasive usually consisted simply of silicon carbide bonded with clay. Additions of high-alumina material such as corundum were sometimes made. More recently, however, silicon nitride has been introduced into the composition. This, together with changes in processing and with such materials as self-bonded silicon carbide, where some, at least, of the material has been produced by reaction *in situ*, has meant that a simple elemental analysis, such as can be achieved using XRF and/or traditional 'wet' chemical methods, provides only a part of the required data. In fact, elemental analysis is often of relatively minor importance to the technologist as compared to a knowledge of what *compounds* are present and their abundance. One of the latest chemical approaches to such a 'rational' analysis for silicon carbide materials has recently been published.[3]

Nevertheless, elemental informaton is still required, as the presence or absence of some elements affects performance. XRF elemental analysis of such materials provides the analyst with many additional problems as compared with the analysis of oxide materials. Not the least of these is the interpretation of the data when it is finally obtained. A more detailed consideration of some of these problems is given in the appropriate sections when discussing the handling of individual types of material. Here, it is sufficient to consider those problems posed by the preparation of a melt suitable for presentation to the instrument.

The completed analysis, expressed as oxides, will resemble closely a high-silica material, and the material will generally be ultimately analysed using the high-silica calibration. This calibration is based on the use of a weight of 1.5 g of sample, mainly SiO_2. However, if 1.5 g of SiC or Si_3N_4 is taken, when the fusion process is completed, the carbon or nitrogen will have been 'burnt out', equivalent amounts of oxygen taking their place. This will have increased the weight of 'sample' in the beads. 1 g of 'pure' silicon carbide will produce almost exactly 1.5 g of silica. Thus, if the calibration is to be that for silica, then it is clear that a lower weight of the non-oxide material, namely, that calculated to yield 1.5 g of oxidized sample, needs to be used. When the nature and probable composition of the sample are known, this calculation is best carried out prior to starting the fusion. However, it not infrequently happens that this information is not available, and even occasionally is given incorrectly.

The range of composition of materials containing non-oxide materials, particularly silicon carbide, is now so wide that a universal approach is not possible. Sample types may include aluminosilicate or higher alumina content materials containing a small amount of silicon carbide, 'pure' silicon carbide, and complex mixtures containing graphite, silicon carbide, silicon nitride, silicon 'metal' and various silicon oxy-nitrides. It is thus possible only to lay down general guidelines designed to involve the minimum effort to achieve a satisfactory result. The use of XRD may help to characterize the species present in a given sample, but much work remains to be done in this area.

When graphite is present, it is essential to burn this out and to obtain a loss on ignition figure (if possible). This is often required as part of the final data, even if it is used only as an easily obtained estimate of the graphite content. It is, however, not always possible to obtain. Ignitions are best carried out at a lower than normal temperature, i.e. 800°C. Even this temperature is an uneasy compromise, as it is difficult to burn out all the graphite, which may be coated with silicon carbide produced during processing of the product. On the other hand, silicon carbide may start to oxidize at about this temperature. As the process of burning out graphite is very slow, often requiring overnight ignition, the effect of even slow oxidation of the silicon carbide can be significant. Many materials respond satisfactorily to this treatment, but the silicon carbide in a few materials decomposes more rapidly, in some cases even to the point where a loss on ignition cannot be undertaken with any hope of accuracy. Used refractories may well contain vanadium pentoxide and nickel oxide (from the use of fuel oil); the presence of these oxides seems to result in a more rapid decomposition of the silicon carbide and at lower temperatures. In some fortunate instances it may be possible to derive a loss on ignition figure adequate for use by measuring the rate of oxidation of the silicon carbide (after apparently burning out the carbon) by successive weighings at intervals and then extrapolating back to zero time. This clearly is a last resort, but may provide some information when any alternative is lacking.

Methods of handling the fusion also vary depending on the data available and the behaviour of the material during the analysis. If sufficient is known about the composition of the material, a weight can be calculated to yield ≃ 1.5 g of oxidized material. If this calculation proves to be sufficiently correct to allow melt weight factors to be used without significant loss of accuracy, this is the course to pursue. The fusion process should be conducted using this calculated weight. Where information is lacking, it is necessary to proceed with a 'trial' fusion. Whereas weighing the melt from this gives some indication of the accuracy of the assumed composition, running the bead on the spectrometer to derive an approximate analysis is also desirable. From these results it is usually possible to calculate a probable analysis. Then, depending on circumstances, it may be possible (having retained the rest of the melt in its dish) to add an appropriate amount of extra flux or sample to the residue and the bead (returned to the dish), re-melt with swirling and re-cast. Better practice, however, is to prepare a second bead after having calculated the correct weight to be used.

The need to oxidize the sample is responsible for the other difficulties in analysing these materials. Decomposition of the material is slow and difficult. Silica reacts readily with the alkaline flux used, but this reaction cannot proceed until silicon, whether elemental or combined as silicon carbide/nitride, has been oxidized. In fact, both reactions appear to proceed virtually simultaneously, oxidation being followed almost instantaneously by dissolution of the silica. Attempts to speed up the decomposition by increasing the temperature of attack are fatal. Platinum and its alloys are very prone to the attack of carbon and especially, it would appear from experience, of silicon. Any raising of the temperature which might allow the molten flux containing undecomposed sample to come into contact with the metal of the fusion vessel can severely damage the dish. At best, and even with care, prolonged use of dishes for fusing reduced samples will result in crystal growth and severe roughening of the surface. The life of the dish is inevitably very much shorter than when it is used for handling normal samples. The presence of particles of silicon arising unexpectedly in a material can produce either pinholes if there are only a few, but significant amounts distributed through the whole can produce a cut through the platinum alloy at the melt/air interface almost as clean as if it were produced by a pair of scissors. When silicon 'metal' is known, or suspected, to be present, even greater than usual care needs to be taken. It may be worth noting here that fused silica, if it has been prepared using graphite electrodes, often contains small grains of silicon that will quickly pinhole dishes.

As yet, no completely satisfactory technique has been developed for decomposing many of these materials. Some of them, such as silicon nitride, silicon oxynitride and the range of sialons (silicon nitride and alumina compounds), are relatively easy to decompose, at least by comparison with silicon carbide. Again, if the material (as in the case of silicon itself) is susceptible to attack by hydrofluoric acid in the presence of nitric acid, it is best attacked in this manner; the residue is analysed by adding 1.5 g of pure silica and then fusing as a high-silica sample. (This approach is, of course, only viable if a figure for 'silica' is not required.) Unfortunately, many of the other reduced forms of silica are not amenable to the attack of hydrofluoric acid. In any event, the effective content of forms of silicon in the material is usually required information, so that a figure for SiO_2 is needed. In the case of the analysis of silicon metal, however, the interest normally lies with the impurities.

In view of the grave danger to the platinum from non-oxidized elements it is imperative to prevent, whenever possible, any contact between the undecomposed sample and the platinum alloy. It is, in fact, the triple contact, platinum alloy, undecomposed material and molten flux, that seems necessary to avoid. The most successful way of doing this so far found is to use the 4:1 flux composition but derived in a different way.

The procedure utilizes the fact that lithium tetraborate and boric oxide have significantly higher melting points than the mixed flux. By means of a simple calculation, the weight of borate in 7.5 g of the 4:1 flux is found; this is then re-calculated as lithium tetraborate and this weight of the latter material is melted into the dish. While the material is molten the dish is swirled so as to provide a layer of tetraborate over that part of the surface (base and sides) likely to be in contact with the melt during the fusion, and the dish is allowed to cool. The amount of lithia present in the tetraborate is calculated to obtain the level of deficiency in terms of that oxide as compared with the use of 4:1 flux. This amount of lithia is then added to the dish as lithium carbonate and intimately mixed with the sample in the dish, in that part of the bowl covered with the lithium tetraborate layer. In some cases boric oxide may be used in place of the tetraborate, with appropriate adjustments to the lithium carbonate amount.

By this means it is intended that the actual decomposition will be achieved in the form of a sinter of the lithium carbonate and the sample rather than a melt. This allows the maximum access of air to provide the necessary oxygen but not contact with the metal of the dish. Great care is required in the early stages of the fusion. Low temperatures, as provided by a small gas burner with the flame being progressively increased, and localized heating under the control of the analyst are desirable. The reaction of lithium carbonate, oxygen, and silicon carbide or nitride is very exothermic, so that personal attention is essential to avoid it running away and raising the temperture sufficiently to penetrate the solidified layer of tetraborate at this early stage. The exothermic nature of the reaction tends to melt areas of the sinter, a process assisted by the fact that the reaction products, lithium silicates possibly mixed with some excess lithium carbonate, have a lower melting point than the original mixture. The exothermic nature of the process is self-evident, as individual particles of sample generate tiny sparks as they react. Heating from above by a gas burner, with careful visual control in these early stages until most of the material has reacted with the carbonate flux, can often achieve relatively rapid decompositions without necessarily attacking the dish.

As will have been realized from the above, the process is not easy, demanding both skill and a high level of attention, possibly for some considerable time. Fusion times may vary from a few minutes to (more likely) several hours. Attempts to 'short-cut' the process will almost inevitably result in damage to the dish. However, once the stage of decomposition in lithium carbonate has been carried as far as appears possible, the remaining stages are relatively risk-free, when at higher temperatures the tetraborate is melted and the sample sinter is incorporated into the whole. Even this process may be somewhat prolonged, as small traces of what appears to be undecomposed sample (but may well be traces of graphite) are often seen floating on the surface of the melt and have to be incorporated before removing the disk from the burners to the furnace for final heat treatment and casting. However, the danger of damaging the dish appears to be confined to the earlier stages of fusion.

It has been noted that, on occasion, iron (and possibly tungsten) is lost to the dish during the fusion. This is almost certainly due to the failure to maintain an oxidizing atmosphere universally through the reacting mass during the whole of the process. It also denotes that some of the molten mass has been in contact with the metal of the dish by penetrating the tetraborate layer. While it may be said that this is a failure on the part of the analyst, it is nevertheless almost impossible to avoid such occurrences entirely. To ensure a correct analysis, particularly with regard to iron, it is very desirable to carry out the analysis in duplicate and to compare the two iron oxide figures. Failure to achieve adequate duplication demands a third fusion. Any iron entering the dish can be detected by heating in an oxidizing atmosphere; a blue discoloration indicates the presence of iron. The method of removal of the iron has already been dealt with (see Chapter 2, Section 2.7.2).

It is appreciated that the present method leaves much to be desired, but it can achieve results. Attempts have been made to carry out these fusions in an atmosphere of oxygen to assist oxidation. Slight improvements were noted, but not sufficient to outweigh the greater manipulative difficulties encountered with the additional equipment. Nitrates are often recommended to assist such oxidatons, but little advantage seems to be gained. There are always dangers in the use of nitrates; firstly, they are hygroscopic and a rapid release of water can cause spurting, second is their well-known tendency to damage the dish.

5.7 CHROME-BEARING MATERIALS

Fusing materials containing medium to high contents ($>10\%$) of chromium sesquioxide (Cr_2O_3) presents considerable difficulties. Relatively small amounts of chrome are met with in pottery glazes and coloured glasses, and these are generally low enough to be catered for by the normal fusion technique. Materials such as vitreous enamels or welding fluxes can be similar, but on occasion they can contain enough chrome to give rise to problems. In many of the above sample types the chromium oxide may, in any event, be effectively pre-fused during the processing of the product. Problems may be expected when analysing samples from the refractories industries. Chrome ore (chromite) is a commonly used additive to magnesite, but in materials described as magnesites it is usually below the level that causes real difficulties. Older type and currently lesser-used materials such as chrome-magnesite and magnesite-chrome refractories, together with the ores themselves, offer major difficulties.

Attempts to decompose the latter at 5:1 or even 10:1 flux to sample ratios using lithium borate fluxes are doomed to failure, ratios as high as 24:1 being required. Clearly this dilution produces problems with regard to sensitivity. Fortunately, the usual technology of these materials shows no interest in alkali contents, otherwise sodium would pose an impossible problem. Even so, results for magnesium show very poor reproducibilities and can be of doubtful value.

These problems would seem to arise because chrome ore consists primarily of various spinels; those combining the oxides of magnesium, aluminium, chromium and iron (both ferrous and ferric) may be present. These are very stable compounds and are therefore very difficult to decompose, even when a free choice of flux is possible. In the standard chemical method for chrome-bearing materials a powerful mixture of sodium/potassium carbonate/borate is used; even there the decomposition ratio is at least 10:1. Nevertheless, this potent mixture can still take about 2 hours to decompose 1.0 g of sample. This decomposition utilizes the conversion of Cr^{3+} to Cr^{6+} by air oxidation, thereby allowing its combination with the alkali in the flux as chromate. This reaction alone is a powerful factor in the decomposition of chrome-bearing materials. Unfortunately, it cannot be achieved using the less basic lithium salts; beads from chrome-bearing samples are green, whereas sodium fusions produce bright yellow melts. Even if dichromates (e.g. that of potassium) are fused in lithium borate fluxes, they reduce to the Cr^{3+} state. In decomposing chromite or its products, there is also the major problem of oxidizing the considerable amount of ferrous oxide in the spinel.

If the hypothesis concerned with acidity/basicity of the sample material is considered, it is not surprising that problems are encountered. The major oxides found in chrome ore and its products are those of chromium, iron, magnesium and aluminium. Chromium, which has just been discussed, is better attacked by an alkaline flux. Iron responds in an uncertain fashion, many analysts believing it does not really go into true solution at all; however, theoretically, an acid flux should be preferable, provided that oxidation to the ferric state can be achieved. The acid flux is required for magnesia and probably for alumina. Summing this up, and bearing in mind that chromium oxide is often predominant, it is almost impossible to select a 'best' flux. Even in the earlier days of 'wet' analysis, many recommendations were to be found for the use of a very acid flux, potassium bisulphate or pyrosulphate, as an alternative to the more frequently used alkaline flux of sodium carbonate/borate. This never appeared very practical, even for wet methods, and does not seem to offer any possibility of casting beads. In practice, success can only be achieved by electing to decompose the sample in a large weight of flux. Even so, the process is protracted, with the material very reluctant to decompose; a considerable amount of patience is required.

It also seems that temperatures of about 950°C, attained over a gas burner, appear to be optimal. This may be because of the tendency of ferric oxide to lose oxygen at higher temperatures. It is important to ensure free access of air to the melt to ensure oxidation of the ferrous iron in the spinel. It is equally important to prevent unburnt gas getting inside the dish, as this can result in reduction of the chromium in the sample, producing dark shiny flakes inside the melt. This is not a common occurrence but, when it happens, the melt is best discarded, as no amount of subsequent heating appears to be able to restore the situation. These flakes appear to contain virtually all the chromium in the sample, as the colour disappears from the melt. This phenomenon seems to be a more frequent occurrence using the wet chemical method of decomposition.

Some analysts within the XRF Working Group reported successful decomposition at a very high temperature, 1400°C, but not only is this difficult to achieve in a general laboratory, but questions also arise about volatilization of both the melt and the components of the sample.

Other possible fluxes have been tried. A number of experiments were conducted with lithium phosphate fluxes, which had been regarded as potentially successful. Marginal improvements in flux/sample ratio were achieved (14/1) using a mixed flux of 9.3 g of Li_3PO_4 and 0.5 g of $Li_2B_4O_7$ to fuse 0.7 g of sample. No improvement in analytical performance was noticeable from the greater concentration of sample, probably mainly because of the higher mass-absorption coefficient of the matrix. Increased interference effects were also observed, so that the use of phosphate fluxes offered no advantages.

Reverting to the discussion on the use of more alkaline fluxes to aid the decomposition by encouraging the oxidation of Cr^{3+} to Cr^{6+}, there seems to be a possibility that the use of even more alkaline oxides than sodium or potassium oxide might help, and possibly permit the determination of the two alkalis of analytical interest. It might seem to be advantageous to attempt decompositions using either rubidium or caesium fluxes with varying alkali to borate ratios, although both oxides are very volatile, which would create problems. In practice these options are not open, as both borates 'creep' when molten and cannot therefore be satisfactorily handled in current types of alloy dishes.

Alternatively, it may even be possible to use the components of the fluxes successively, e.g. using first alkali carbonate, with or without a minor addition of borate, until attack appears to stop. This could be followed by further heating with the remaining borate added. Alternatively, it would be possible to try an acid attack with boric acid first, again with or without a small amount of the alkali carbonate, followed by the remainder of the carbonate later. The first of these two options would probably be the more successful with higher levels of chrome, while the converse might be true where the ore has been admixed with large quantities of magnesia. A technique similar to the second approach has been applied to the analysis of magnesite by direct-reading spectrometry, with considerable success. Chrome ores, however, are so difficult to attack that success is questionable.

A final problem remains. Melts containing significant amounts of chromium tend to stick to the casting dish, giving the beads an enhanced tendency to crack on cooling. This same problem is met with a number of other elements, and is dealt with under the next heading.

5.8 PROBLEM ELEMENTS

There are a number of elements that give rise to problems when they are present in the sample. In most cases, small amounts can be tolerated without cause for concern, but as the level of content rises difficulties will be encountered. These problems can arise in a variety of ways, but they can be characterized under three headings.

(1) *Reducible elements*. These are elements liable to be reduced during fusion and then to alloy with the platinum/gold of the dish.

(2) *Volatile elements*. These are elements that tend to volatilize during fusion, particularly from some matrices.

(3) *Concentration limited elements*. These are elements capable of being brought into the melt up to only a limited level of content.

The three types are not mutually exclusive; in fact, in many cases elements fall into category (3) merely because they are in category (1) or (2). Elements giving rise to these problems are discussed briefly here, but greater detail will be found later.

5.8.1 Reducible Elements

Many of the heavy elements and the transition elements are easily reduced and, once reduced, will find a home in the platinum/gold of the dish. This not only destroys the validity of the analysis but also can affect the next sample in the dish, as well as causing serious damage to the dish. Mention has already been made of the problems associated with the presence of iron. Iron can, in fact, be considered a marginal case, in that a reducing atmosphere or situation is usually needed before reduction takes place. At lower temperatures this is true of some other elements, a very important instance being that of lead. Below about 1000°C lead does not appear to auto-reduce but, even so, great care has to be taken to ensure the total absence of reducing agents. Burner gas that has not been fully burnt will readily reduce lead, as will traces of carbon or carbonaceous matter in the sample. In this connection, it needs to be noted that most glazes used in the pottery industry will have some inclusion of organic materials, possibly as dispersing agents. Similar conditions may well occur in vitreous enamels and welding fluxes. Also, the actual process of grinding, if carried out in a mechanical grinder such as a TEMA mill, will probably have been done using a tungsten carbide vial. The carbide contamination can therefore act as a focal point for reduction. Even with ready-ground samples there is always a possibility that the commercial grinding process has involved the use of, say, a rubber-lined mill. It is clear that the greatest of care should be taken when heating the sample/flux mixture so as to ensure that the mix stays 'open' to allow the burning off of all organic matter or carbon before reaching dangerous temperatures. It is unfortunate that it is usually these samples that have to be fused in their unignited state because they are designed to have relatively low melting or softening points. Thus, the sample tends to sinter, if not melt, during the determination of loss on ignition; this in turn prevents the fusion sample being taken from the ignited material and necessitates the use of a raw sample for the fusion.

The same situation can arise with many of the elements used in glazes, special glasses, vitreous enamels and welding fluxes, and also with slags, etc. Some smelting and other processes may involve the same group of elements. Danger can easily arise from materials containing, for example, arsenic, antimony, copper, cobalt and even zinc. Slags, of course, are very liable to contain metal, which poses its own problems, not least in regards to damage to the platinum.

It will also be noticed that many of these elements that can be reduced also tend to stick to the fusion and casting dishes. The former is of less importance unless it reaches such proportions that significant amounts of the melt cannot be poured into the casting mould. Small weights of bead result in a larger radius of curvature in the bead, i.e. a flatter bead. This, as has already been noted, leads to lower count-rates and lower apparent contents.

There is also a much greater tendency to stick to the casting mould, resulting in difficulties in releasing the beads and a marked increase in the proportion of cracked beads and, thus, the number of re-melts. This problem can be very considerably reduced by the addition either to the melt or to the heated mould, immediately before pouring in the melt, of a crystal of lithium iodide or iodate. Some analysts have made a practice of using the bromide, but experience shows that the iodide or iodate acts as a more effective releasing agent and, in any event, bromine interferes with the determination of aluminium. The weight of lithium salt added should be small so as not to disturb significantly the sample/flux ratio.

Many of these same elements tend, because of the thermodynamics (ΔG values) of the oxide-oxygen to metal balance ($M_xO = xM + O$), to lose oxygen simply as a result of heating to high temperatures. This difficulty is clear-cut with lead and cobalt, neither of which can be guaranteed to be retained in the melt at 1200°C. Both will reduce readily if they are present in anything but small amounts and, even when the level of content is low, care needs to be taken. Lead is capable of causing severe damage to the fusion dish, to such an extent that it will almost certainly have to be sent back for replacement; a very expensive process. It is fortunate that almost all materials containing lead in significant amounts can be fused readily in 4:1 flux at 1050°C without difficulty—but with care!

5.8.2 Volatile Elements

Among the above elements, arsenic and antimony are mentioned as elements that tend to reduce, but more importantly they tend to volatilize. To these could be added cadmium, which under certain circumstances will volatilize unless special care is taken. The problem with elements that tend to volatilize is that it is very difficult to prevent such occurrence except by keeping down the fusion temperature. It is necessary to investigate the conditions under which retention of all the elements present can be guaranteed during fusion. There are so many factors that tend to affect the onset of and the degree of volatilization that it is almost essential to conduct appropriate experiments within one's own laboratory. Experience has shown that, with care, cadmium and arsenic can usually be retained. The greater acidity of lithium borate fluxes as compared with carbonate/borate mixtures of sodium and potassium probably reduces this problem in XRF work.

Strangely enough, it has proved possible to retain such volatile oxides as sulphur trioxide in the bead under favourable circumstances. Gypsum and plaster are commonly used materials in the pottery industry, and the determination of sulphate by traditional methods is very time consuming and therefore expensive. By keeping down the time and temperature of fusion it has proved possible to retain, within experimental error (say $\pm 0.25\%$), the whole of the sulphur trioxide in the sample, and to determine it against a range of standards prepared in a similar fashion. As has already been discussed, one would naturally expect sulphur trioxide to be retained more readily in a metaborate-rich flux than one consisting mainly of tetraborate. Similarly, it is to be expected that SO_3 can be more reliably retained and determined in the presence of mainly alkaline oxides, e.g. CaO, Na_2O, etc., than in silicates or aluminosilicates. Sulphur present as sulphide is almost impossible to retain quantitatively unless it can be first oxidized and 'fixed' as sulphate.

5.8.3 Concentration Limited Elements

Tin is an element used widely in the pottery industry in glazes as an 'opacifier', namely, for its ability to reduce the transparency of the glaze, giving a whiter appearance to the ware. It was used previously in the sanitary ware industry as

DECOMPOSITION OF SAMPLES BY FUSION   91

an 'engobe' or intermediate layer between the cane or buff coloured fireclay body and the glaze. The virtual disappearance of sanitary fireclay before the onslaught of vitreous china, a pure white body, has largely removed the need for this additional layer. Even as an opacifier in the glaze, tin has been largely replaced by a less expensive combination of zinc and zirconium oxides, but it may still be used on occasion. Tin functions so effectively as an opacifier because it acts as a nucleating agent and refuses almost completely to dissolve in the glaze. This technological advantage, as in so many other cases, proves to be a grave disadvantage to the analyst. Exactly the same phenomenon arises when the attempt is made to bring stannic oxide into the flux. It is almost impossible to dissolve, remaining as discrete particles in the glaze once the content rises over a low level (approximately 10%). Dilution is again the only answer.

Copper has been mentioned above as a reducing oxide. Although this is correct it is not the main problem. Copper forms melts that are very difficult to part from either the fusion or the casting dish. To overcome this problem, dilution has to be effected to such a degree that the problem from reduction is almost eliminated, at least to the point of being of little significance. (It is not unreasonable to hypothesize that the tendency to stick to the dish and mould is related to a tendency to reduce to the metal and form alloys). An upper limit for viable operation would be about 3% CuO content in samples fused at the normal 5:1 dilution.

5.9 DILUTION

Dilution, i.e. increasing the flux to sample ratio, has been mentioned as a solution to several of the problems above. Nevertheless, dilution inevitably leads to some loss of sensitivity, even if not in strict proportion. Dilution may be effected in two ways. On the one hand, calibrations may be made using higher levels of dilution than 5:1 to allow a particular class of materials to be analysed, e.g. chrome ores. On the other hand, it not infrequently happens that the sample will fit a particular calibration type except for a high level of a concentration limited element, say copper. In such a case it may be advantageous to make two beads, both with an appropriately lower weight of sample, say 0.5 g, and make the weight of 'sample' up to 1.5 g by the addition of, in the one case, silica and in the other alumina. This gives a dilution factor of 3, beads that can be run on an existing calibration, and two sets of results from which the composition of the original material can be calculated.

When the dilution factor exceeds about 5, sensitivity and thus reproducibility of the analysis may drop below acceptable levels. However, in many of the cases where dilution to such a degree proves necessary, methods of accurate analysis of the material concerned are difficult by any other technique. Whether 'wet' methods or techniques such as AAS or ICP are used, accuracies and precisions obtainable are usually no better and often worse than by XRF, even with dilution. The ease with which XRF analysis can be achieved usually means much lower costs compared with most other techniques. The greater speed, simplicity and cost of XRF analysis frequently tips the scales in favour of its use for these types of complex materials. In any event, the problem of loss of sensitivity may often be ameliorated by carrying out multiple fusions using different diluting oxides, longer counting times or, better, both. If several beads are fused and run, the counting statistics will improve by one over the square root of n, where n is the number of beads run.

The use of 'wet' methods is often precluded by the almost virtual impossibility of some of the separations involved. Typical examples that occur frequently are the separation of elements often found together in glazes and vitreous enamels, such as lead, tin, arsenic and antimony. Again, in vitreous enamels one finds together cobalt, nickel, manganese and zinc, which are almost impossible to determine accurately in such a mixture by chemical methods. A further example, from either nature or man-made sources, is the separation of calcium, barium and strontium. Only since the advent of XRF has it become feasible to accept on a routine basis, at least to industrial analysts, that nature rarely provides one of these elements without the presence of small amounts of the others. Gravimetric (oxalate or sulphate) or volumetric (EDTA) methods invariably give some form of summation of the total contents, a summation which is normally incorrect. Gravimetrically, results tend to be in error owing to incomplete precipitation of two out of the three, and volumetrically the standardization factor can only be correct for the element sought.

Even 'specific' techniques such as AAS and ICP are subject to interferences, for example, the effect of aluminium on the alkaline earths in AAS and its background effects below 230 nm in ICP. Provided that the analyst is carrying out control analyses on a repeating source of raw material or product, these interferences can be evaluated at a cost that will be repaid as the number of analyses using the procedure increases. However, where the laboratory is concerned with a wide variety of materials of different composition within a given type, the investigation and evaluation of such interferences may become uneconomic, XRF once again scores, in that interferences are readily evaluated and correction factors are generally applicable, without loss of accuracy, over a wide range of compositions.

The case of chrome-bearing materials has been mentioned above and is one of the instances where a normal, industrially used ore requires some dilution at higher levels of content (say $Cr_2O_3 > 40\%$ when the flux for magnesite/chrome refractories is used). Except in the few instances where a 'pure' prepared chromic oxide is the subject for analysis, dilution factors are not likely to exceed 2. The resulting analyses, even with a standard flux/sample ratio of 24 diluted by 2, still produce better results than are generally obtainable by other methods.

Thus, the issue of the acceptability of XRF analysis when dilutions are involved must be considered in the context of the validity and reliability of the results, making full allowance for the accuracies required and the cost of obtaining them. XRF results, given that calibration and evaluation of interference effects have been properly carried out, should not suffer from significant errors in accuracy, but only in precision: i.e. dilution affects the precision only, except when impurity levels in the flux become significant (e.g. CaO, Na_2O) or where background counts are high. Recognition of achievable accuracies is the main reason why results are often issued quoted only to the first decimal figure for all contents of 10% or over. Even this is questionable, since it would be a bold analyst who would offer a *guarantee* of results better than, say, 0.5% of content, i.e. 1 in 200. Even 1 in 200 is, in reality, probably optimistic at lower levels of content, where absolute, rather than proportional, errors begin to affect the accuracy; then, as the level goes down further, these zero errors dominate.

REFERENCES

1 Oliver, G. J., Hodson, P. T. A. and Sweeney, C. Q. (1986) *Contamination from McCrone Microniser*, BCRL Technical Publication No. 115, British Ceramic Research Ltd, Stoke-on-Trent.
2 Oliver, G. J. (1972) *The Analysis of Aluminosilicates by X-Ray Fluorescence*, Report of the Analysis Committee: XRF Working Group, Special Publication No. 73, British Ceramic Research Ltd, Stoke-on-Trent.
3 Julietti, R. J. and Reeve, B. C. E. (1991) *Br. Ceram. Res. J.*, **90**, 85.

6 Selection of Instrument Parameters

6.1 INTRODUCTION

The theory behind X-ray fluorescence spectrometry is not within the scope of this book; it has been adequately dealt with in several works. If this sort of information needs to be sought, some of the sources are noted at the end of this chapter. The prime aim here is to consider instrumental parameters from the point of view of the particular sorts of sample that the analyst in this field might encounter. Although most of the discussion is applicable to the needs of geochemical analysts, they may regard some of the recommended settings as unusual. This is because in many samples the geochemical analyst will generally encounter elements other than the conventional eight at either trace or lower minor content levels, and will choose settings appropriate to this situation. The ceramic analyst, on the other hand, will encounter many of these other elements as deliberate additions in a wide range of contents from trace levels to almost 100%.

This chapter deals with the choice of X-ray tube and the readily changed parameters affecting the selection of lines. The following chapter discusses the factors affecting the choice of individual element lines.

At this point it should be noted that much of this information relates primarily to the instruments with which the authors have had opportunity to work. Instruments available at the time of writing in the authors' laboratories are a simultaneous Telsec TXRF, an ARL 8480 and a Philips PW 1606. Previous experience has been with a Philips PW 1410 and a simultaneous/sequential Telsec TXRF. Even so, much of the discussion should be relevant to other types of instrument; certainly the same general principles of line selection will apply. In the case of tubes, almost all the experience has been with chromium or rhodium target tubes; this is because most of the interest in rock, mineral and ceramic analysis is directed towards the lighter elements. While the purchase of a new instrument was being considered, tube selection was made on the basis that it was better to pay the price of reduced sensitivity for the heavier elements, particularly as there was little doubt that it would still be adequate for most purposes, in order to gain better sensitivity for the lighter elements. Even so, this is less than might be desired in some instances. This assessment has been fully justified by the results achieved with the rhodium tube.

6.2 X-RAY TUBES

The ceramic and the geochemical analyst could well differ about the best choice of X-ray tube. This is because of their differing requirements. Often the geochemical analyst needs to provide the best possible information almost regardless of cost. This is exemplified by the use, where necessary, of extremely long counting times for both line and background in order to achieve accurate determinations of trace elements. The cost, including even the down-time involved in changeover of tubes, is often acceptable if better data results. This is not an option for the ceramic analyst; the same tube *must* remain in place throughout its effective working life. It is very rarely that the data provided by a comparatively short counting time (100 s or less) is not of sufficient quality to satisfy the needs of the ceramic technologist. In those few cases where this is not so, the cost of improving the quality of the information is often prohibitive and alternative techniques will be sought. It is almost impossible to visualize circumstances where the ceramic analyst could contemplate the cost of a tube change as a semi-routine procedure. Its effect on the time taken for and the cost of the analysis in question, to say nothing of the delay engendered for every subsequent analysis, would certainly not be acceptable.

The above means that the original choice of tube is of great consequence to the ceramic analyst, as it must be able to yield adequate sensitivities over the whole range of potentially required determinations. In the past, the chromium target tube was used almost exclusively with considerable success, but it is now being progressively replaced by rhodium and scandium tubes. At the time of writing, dual target tubes, both dual-coated and segmented, are making their appearance, but it is too early to make any comment about their potential value.

The chromium tube has good sensitivity for light elements, but has an embarrassingly high sensitivity for lines such as calcium $K\alpha$, potassium $K\alpha$, barium $L\alpha$ etc., resulting in two problems. Second or third order interferences occur which cause problems with the determination of the lighter elements. The most significant of these are a third order calcium $K\alpha$ line affecting magnesium $K\alpha$, a third order barium $L\alpha$ line affecting aluminium $K\alpha$, and a second order zirconium $L\alpha$ line affecting sodium $K\alpha$. A second problem is that for high concentrations of some of these heavier elements, even at the dilution in a fused, cast bead, the count-rate is too high if the same tube power is used as may be needed for other elements. This applies particularly to the determination of CaO in cement, limestone, dolomite and bone ash, when either very low power settings must be used or recourse must be made to the $K\beta$ line. The problem of high count-rate is more serious with older instruments, because the range of linearity of detectors is lower.

The rhodium target tube yields a much flatter response curve throughout its excitation range, and the problems of second order interference are therefore

SELECTION OF INSTRUMENT PARAMETERS 97

less significant. This enables the same power settings to be used across the whole range of determinations without penalty, making it ideal for use with a simultaneous instrument for this sort of work.

A further problem arises with the chromium target tube, in that chromium is, of course, a frequently required determination in ceramic materials. Rhodium, on the other hand, occurs very rarely, and then only in very specialized materials, such as are usually unsuitable for XRF analysis. The problems created by having a target of the same element as is required in the analysis are obvious. Chromium target tubes have also been found to give rise to difficulties in the determination of manganese, praseodymium and europium. The first of these is relatively common, the second occurs as a constituent of some glazes, glasses and colours, the last is, to say the least, uncommon. The only element so far encountered that appears to give problems with the rhodium tube is the rarely required ruthenium. If necessary, rhodium and ruthenium can be measured by using a molybdenum filter in front of the tube.

A rhodium tube offers better sensitivities for shorter wavelengths and produces a useful Compton peak of value for the calculation of mass-absorption and background correction in the semi-quantitative analysis of powdered samples.

Scandium target tubes are, at the time of writing, too new to be available for all types of instruments, notably simultaneous spectrometers. Nevertheless, their value as compared with rhodium tubes is of interest, as the option is already open to many users and will become more widespread as time goes on. Rhodium tubes have advantages for simultaneous instruments. Comparative trials indicate that scandium has a better sensitivity for the light elements such as fluorine, sodium, magnesium, aluminium, silicon and phosphorus. It also shows a very marked increase in sensitivity for calcium and potassium but worse sensitivity for the short wavelengths. It also causes problems with high order reflections on light elements (though less than chromium) and, of course, is worse for the determination of scandium, a not very common requirement. On the other hand, because scandium tubes are normally of the side-window type, they cannot be sited as close to the sample as an end-window tube. This means that the response to a rhodium end-window tube is very similar to that to a scandium side-window tube for light elements. On balance, it may be concluded that the rhodium tube has the edge overall unless high sensitivity for the light elements is really critical.

6.3 X-RAY TUBE PARAMETERS

The settings recommended by the manufacturers for a chromium target side-window X-ray tube included power settings of 45 kV, 60 mA for 'light' elements. This was for use up to the K $K\alpha$ line, above which for the 'heavy' elements the recommendation was 60 kV, 40 mA (Ti $K\alpha$ and above). In practice, power settings for all but low levels of lime had to be reduced because of its embarrassingly high sensitivity. These general parameters served well for the life of the Philips PW 1410.

The simultaneous Telsec TXRF was equipped with a rhodium target end-window tube. For many years, in order to achieve maximum sensitivity, it was operated at its maximum power setting of 50 kV, 50 mA. This gave a good compromise of sensitivity for those element lines in the wavelength range 1.37–11.9 ångstrom. Later it was found by bitter and costly experience that, contrary to recommendation, the generator should only be run continuously at a total power rating of 2 kW. This, of course, necessitated a power reduction by decreasing mA but not kV for subsequent activities, coupled with a 25% increase in integration times.

The ARL 8480 equipped with a rhodium target end-window tube, being a sequential instrument, opened up possibilities of using varying power parameters so as to provide optimum conditions for each element. The general rule to achieve best sensitivity is to operate the spectrometer at maximum tolerable power. Most modern instruments do not permit infinitely variable parameters, and both voltage and current are usually in steps of 5 or even 10 kV or mA. This can mean diverging from the optimum in some cases so as to use the more important maximum power settings. For example, if the maximum setting for sensitivity for a given element is at 55 kV, neither 55 nor 60 mA could be used as these would exceed the power maximum of the instrument. This would mean using only 50 mA and thus a total power of 2.75 kW, whereas better results would be achieved using settings of 50 kV/60 mA, allowing the full 3 kW to be used. The ARL 8480 used by the authors utilizes a low kV generator, as the alternative high kV generator is not a viable tool for ceramic analysis. This gives possible extreme settings, utilizing the full power of 30 kV/100 mA and 60 kV/50 mA. The flexibility of this instrument and the more recently acquired Philips PW 1606 has enabled the whole question of isowatt kV/mA ratios to be studied at some length.

Utilizing the two extremes (30 kV/100 mA and 60 kV/50 mA), the relative sensitivities for a number of typical lines are tabulated below for commonly analysed elements. In Table 6.1 the abbreviations I_{30} and I_{60} are used for the intensities produced by the use of the 30 kV and 60 kV settings respectively.

From these figures (Table 6.1), it is clear that the balance of advantage crosses over between V $K\alpha$ and Cr $K\alpha$ and, by implication, between Ce $L\alpha$ and Nd $L\alpha$.

Table 6.1. Relative sensitivities at 30 kV/100 mA and 60 kV/50 mA

Line	Ca $K\alpha$	Sc $K\alpha$	Ti $K\alpha$	V $K\alpha$	Cr $K\alpha$	Mn $K\alpha$	Fe $K\alpha$	W $L\alpha$	Zn $K\alpha$	Hf $L\beta$
Energy (kV)	3.67	4.09	4.51	4.95	5.41	5.90	6.40	8.33	8.64	9.02
I_{30}/I_{60}	1.23	1.18	1.08	1.02	0.94	0.86	0.84	0.59	0.62	0.56

From this it may be concluded that, for 12 of the 19 most commonly used lines, the 30 kV setting is to be preferred. With the exception of Hf $L\beta$ and W $L\alpha$, more than adequate sensitivity is obtained even at 30 kV. With reference to the latter, tungsten is normally only monitored to correct for minor levels of contamination, so that ultimate sensitivity is not very critical. In addition, Hf $L\beta$ benefits from the use of the low kV setting because this substantially reduces the line overlap deriving from Zr $K\beta$ (by a factor of about 4). Therefore, in simultaneous operation, there is much to be said for using the lower 30 kV setting throughout. Even in sequential operation it takes the generator on the ARL 8480 instrument about 45 s to change settings; this is potential analysis time lost. This time might well be more gainfully employed to recoup sensitivity on the higher energy lines by using longer integration times with a single power setting throughout. Thus, unless analytical programmes call for the determination of a significant number of high energy lines (e.g. trace element levels of PbO, As_2O_3, ThO_2 and U_3O_8), the best overall operation is provided by the use of a single low kV setting.

The Philips PW 1606, when operated at maximum power, has a minimum voltage setting of 40 kV. This is the ideal setting for most ceramic and similar applications, and provides the advantage of better sensitivities for elements such as titanium, chromium, manganese, iron, tungsten, zinc and hafnium than a 30 kV setting (viz, ARL 8480).

A study was carried out to determine the optimim kV setting for a whole series of lines of ceramic interest. The Philips PW 1606 allowed several lines to be studied simultaneously between 40 and 60 kV, and the ARL 8480 allowed lines to be studied singly down to 20 kV. Plots of intensity and kV at isowatt conditions appear as parabolic curves; these allow peaks to be estimated to within 5 kV in the range 20–60 kV. The results of these trials are depicted in Fig. 6.1, which shows the relationship between the energy of the excited X-ray and the kV setting that gives maximum intensity. The values for Mn $K\alpha$ and Fe $K\alpha$ are a little less accurate than most because their preparation involved extrapolation above 60 kV. The conclusion is readily drawn that those lines that are excited by Rh $L\alpha$ radiation all require an optimum voltage of $\simeq 27$ kV. This appears to be true irrespective of whether they are $K\alpha$ lines (Na to Cl), $L\alpha$ lines (e.g. Zr) or even $M\alpha$ (e.g. Pb) lines. Measurement of the back-scattered Rh $L\alpha$ line showed that it peaked only a few volts above this value. From Ca $K\alpha$ to Fe $K\alpha$ the optimum kV setting increases in proportion with the energy (E_c) of the line itself, although the relationship does not pass through zero. The following equation appears to relate line energy to voltage for maximum intensity:

$$V = K(E_c - E_{Rh})$$

where $K \simeq 23$. Little is therefore to be gained by using voltage settings below 30 kV.

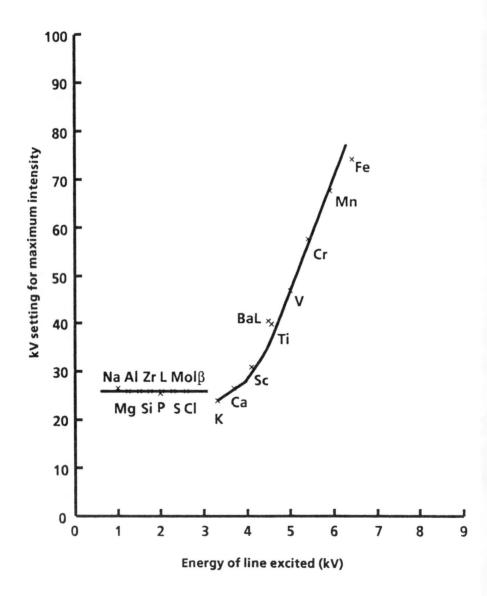

Fig. 6.1. kV settings for maximum intensities

SELECTION OF INSTRUMENT PARAMETERS

As sensitivity and low detection limits are more difficult to achieve at long wavelengths than at short ones (there often being both to spare in the latter case), for simultaneous work and even for large sequential programmes 30 kV has proved to be the optimum setting. However, where individual elements or groups of elements are sought that are excited by either Rh $K\alpha$ or bremsstrahlung, then the kV should be set to an optimum value as is dictated by these needs; provided that neither higher order lines nor background provides a critical favourable factor. For much trace element work with the heavier elements 60 kV is to be preferred.

Further information on tubes and excitation conditions is to be found in the excellent paper by Feret,[1] who compared kV settings from 60 kV down to 30 kV for rhodium end-window, rhodium side-window and scandium side-window tubes. He concludes that the rhodium end-window tube is capable of yielding the greatest sensitivity, except for Ca $K\alpha$ and K $K\alpha$ lines; kV settings of 30–40 kV are ideal for elements of atomic number up to 23. His studies on backgrounds show a rapid increase with increasing kV settings for rhodium target tubes but a lower rate for scandium target tubes.

6.4 INSTRUMENT MASKS

The mask of the sample cup is normally made of stainless steel. In order to prevent the mask itself being excited, thereby producing strong iron, chromium, etc., lines, an internal instrument mask is used in the spectrometer. It is situated between the X-ray tube and the sample. Normally this is gold plated on the side nearest the tube (this is also the side nearest the detectors). It has been found that gold causes a number of intereferences throughout the spectrum. Silver appears to be a much better option, as its lines are close to those of rhodium, are fewer in number and do not obscure any important analytical lines. This is the option used on both the Philips PW 1606 and the ARL 8480 in the authors' laboratory.

There is an inherent problem with all internal masks. This is that with time they pick up both silicon and sulphur contamination from inside the instrument. The only solution to this is to mask the primary collimators of the sulphur and silicon fixed channels and the goniometer itself so that the detectors do not 'see' the mask. Some instrument manufactures do this as standard, but it is as well to check this out before purchasing a spectrometer.

6.5 SAMPLE MASKS

As stated above, most instruments limit the primary X-ray beam so that it does not excite the mask, thereby providing the best solution. If this is not the case, pure silver has been shown to be a very useful material and is also much cheaper than gold. Masks made from either of these two materials can be obtained from Engelhard or Johnson–Matthey by special order. An alternative solution is to use a mask made from the same material as the X-ray target (this is not applicable in the case of scandium). At present the authors have not been able to obtain a rhodium mask. The other answer is to have sets of masks made of two or more materials and to use these in pairs as may be appropriate. The choice in any individual instance depends largely on the instrument and the tube.

6.6 PRIMARY COLLIMATOR

In almost all modern systems the analyst usually has the choice of fine or coarse collimation. It is appreciated that the angle of divergence is the parameter that matters and that, for a flat crystal, fine collimation can be attained by narrower spacing of the blades, by a longer collimator or both. Nevertheless, the convention of referring to the terms fine or coarse collimation, as they are normally understood, will be used throughout. After all, the customer has no control of spectrometer design and has to accept those collimation conditions that are available for the particular application. As a general rule, for $K\alpha$ lines coarse collimation is used up to calcium; any exceptions to this rule are discussed under the section on individual elements.

6.7 CRYSTALS

The authors have first hand experience of thallium acid phthalate (TlAP), rubidium acid phthalate (RbAP), ammonium dihydrogen phosphate (ADP), germanium (Ge) pentaerythritol (PE), lithium fluoride (LiF 200 and LiF 220) and indium antimonide (InSb) and the new layered crystals for light elements. For convenience the abbreviated nomenclature will be used subsequently. Both ADP and RbAP have been effectively superseded by TlAP, as the latter is far superior to ADP for magnesium and to RbAP for sodium and fluorine. At the time of writing, layered crystals appear to be superior to TlAP for magnesium, sodium and fluorine, with sensitivities about three to four times higher. On the other hand, they appear to have higher backgrounds, but nevertheless detection limits are better. As the 2θ angles used for layered crystals are lower than for TlAP and the peaks produced are generally broader, line overlap (e.g. Zn on Na) is generally worse. Crystal choices for individual lines will be discussed later, but a general rule of thumb has been to use LiF 200 for any measurement that is not beyond the range of the goniometer, then to use Ge to its maximum range, so to PE and eventually to TlAP (or, in newer instruments, a layered crystal). In cases of severe line overlap LiF 220 is used, or for second order overlap, Ge.

However, the penalty in either case is approximately a three-fold loss in sensitivity (except at high angles for LiF 200).

6.8 DETECTORS

A range of detectors has been used; these include argon/methane and helium/carbon dioxide flow proportional counters, krypton and xenon sealed counters and a scintillation counter. The xenon sealed counter is effectively equivalent to a scintillation counter, as xenon and iodine are close to each other in the Periodic Table, and it is often used in fixed-channel instruments. In practice, its resolution has been found to be about half that of a scintillation counter, and the xenon counter has been known to deteriorate. Xenon also produces more intense escape peaks, so that high order interferences are more of a problem.

As a rough guide, for K lines, flow counters seem to be best for wavelengths longer than or equal to those of titanium $K\alpha$ or Ba $L\alpha$, below which the krypton counter is to be preferred. From the wavelength of vanadium $K\alpha$ downwards, the escape peak causes a problem when a flow counter is used and the lower setting on the pulse-height window has to be carefully placed between the true peak and the escape peak. This results in a loss of about 10% of the sensitivity, but setting the window below the escape peak would introduce 'noise', potentially causing even greater problems. With the krypton counter this is not a problem. In addition, as the wavelength decreases below this region, the sensitivity of the krypton detector increases so as to be better than the alternative. However, the krypton counter does suffer from escape peaks from high order lines notably that of zirconium on manganese $K\alpha$, sufficient to prevent its use for the determination of manganese in materials with a major content of zirconia. At wavelengths shorter than about zinc $K\alpha$, the scintillation counter has the edge in sensitivity. On fixed-channel instruments xenon sealed counters are often used, whereas a scintillation counter would be used on a sequential spectrometer.

There are exceptions to these generalities, but these will be discussed in the context of escape peaks and crystal fluorescence in the next two chapters, especially for hafnium.

6.9 FLOW COUNTER WINDOWS

As with the instruments themselves, the design, manufacture and materials used are not usually amenable to user specification and have to be accepted. The appropriate thickness of window for specific elements is a matter of choice. In general, 2 μm windows have provided the best compromise between longevity and sensitivity for a sequential instrument and light element fixed channels. However, 6 μm windows give adequate sensitivity for wavelengths shorter than that of silicon $K\alpha$ on a fixed channel, and are therefore to be preferred because of their greater robustness and, thus, life in service.

6.10 PULSE-HEIGHT SETTINGS

These are best discussed, where necessary, in the context of individual lines. For newer instruments it is possible to plot pulse-height settings automatically and to choose the optimum from the acquired data. For older instruments without this facility, a threshold of 2 and a window of 8 (on a scale of 10) have been used throughout, with the exception of sodium (threshold 3.5, window 2.5) and magnesium (threshold 2.5, window 3.5). These latter choices have been made because the narrow windows reduce the effect of higher order reflectance interferences and background scatter.

REFERENCE

1 Feret, F (1986). Effect of scandium and rhodium X-ray tube parameters on the analysis of light elements, *Can. J. Spectrosc.*, **31**, 15.

RECOMMENDED READING

1 Jenkins, R. and De Vries, J. L. (1975) *Practical X-ray Spectrometry*, Springer-Verlag, New York.
2 Bertin, E. P. (1970) *Principles and Practice of X-ray Spectrometric Analysis*, Plenum Press, New York and London.
3 Tertian, R. and Claisse, F. (1982) *Principles of Quantitative X-ray Fluorescence Analysis*, Plenum Press, New York and London.
4(a) De Jongh, W. K. (1973) *X-ray Spectrometry*, **2**, 151.
 (b) De Jongh, W. K. (1979) *X-ray Spectrometry*, **8**, 52.

7 Element Line Selection

7.1 INTRODUCTION

The elements to be considered may be conveniently divided into three types according to the probable frequency of the need to determine them as constituents, as follows:

(1) the eight oxides normally accepted as those required to provide a conventional 'full' or 'complete' analysis of most common materials;

(2) other commonly encountered oxides/elements;

(3) oxides/elements that are normally called for infrequently in ceramic or allied materials.

Altogether 51 elements are discussed here. This gives some impression of the current range of ceramic materials and the variety of data demanded about them. The information derived from this chapter should enable a wide range of minerals of interest to the geochemist to be handled by virtually identical techniques. The fact is also highlighted that the demands of modern ceramic analysis could not have been met without techniques such as XRF and, more recently, ICP. This could, in turn, have seriously inhibited modern technological developments.

The elements discussed here are shown on the schematic diagram (Table 7.1). This shows the various element lines against the analytical parameters that would normally be expected to be used (the last four columns should be read vertically). For example, from this diagram it can be seen that the expected parameters for, say, scandium are the $K\alpha$ line with the use of a LiF 200 crystal with a fine collimator, low kV, high mA, together with an argon/methane flow proportional counter. In the following text only deviations from such parameters are discussed in detail, although more information is usually given about the 'normal eight' elements and, possibly about those elements in category (2). Deviations may be found between the text and Table 7.1; these are due to specific problems associated with the lines in question. Reasons are usually given in the text. Some indication is also given of the likely occurrence of each oxide/element prior to the XRF parameters and the normal reporting form. Elements in category (1) are given first, as they are frequently required; the others are placed in order of ascending atomic number. All the rare earths are gathered together where the lanthanides would be placed. Standard parameters as derived from the diagram are given in most cases at the start of each element discussion.

Table 7.1. 'Normal' optimum conditions for elements

Element and line						Detector	Power	Crystal	Collimator
Kα	Kβ	Lα	Lβ	Mα	Mβ				
F						↑	↑	↑	↑
Na								TlAP	
Mg								↓	
Al		Br				A	L	↑	
Si		Rb				r	o	P	C
		Sr				g	w	E	o
		Y				o		↓	a
P		Zr				n	k	↑	r
S		Mo	Pb				V		s
			Bi			+		Ge	e
Cl									
			Th			M	+		
		Cd	U			e		↓	
K				U		t		↑	
		Sn				h	H		
Ca		Sb				a	i		↓
		I				n	g	L	↑
Sc		Cs				e	h	i	
Ti		Ba						F	
		La				↓	m		F
		Ce				↑	A		i
V		Pr					↓	2	n
		Nd	Ce			K	↑	0	e
Cr			Pr			r		0	
		Sm	Nd			y	H		
Mn						p	i		
Fe		Gd	Sm			t	g		F
Co			Gd			o	h		i
Ni		Yb				n			n
Cu		Ta					k		e
Zn		W				↓	V	L	
Ga						↑		i	
Ge							+	F	
		Bi							F
Se						S	L		i
Br	As					c	o	2	n
		Th				i	w	0	e
Rb		U				n		0	
Sr						t			
Y							m		
Zr							A		
Nb						↓	↓	↓	↓

Scint = Scintillation counter.
For hafnium see Section 7.3.30; for selenium see Chapter 3.

ELEMENT LINE SELECTION

Occasionally, comments are made concerning an interference produced *by* the element under discussion *on* other elements. Such comments are placed in italic print to draw attention to the fact that such interferences may need to be guarded against.

Most of the interferences reported in this chapter have been measured on an ARL 8480 spectrometer using the goniometer as opposed to the fixed channels. For this reason, references in the text to the magnitude of interferences give those that have been measured on the ARL 8480 unless otherwise stated. Although the actual magnitudes of line overlaps will vary with the type of spectrometer and X-ray optics, the figures should still prove a guide as to how serious a particular interference is likely to be. It may be worth noting in this context that the ARL 8480 appears to suffer from bigger second and third order interferences than expected or found on other spectrometers. The reasons for this are not clear, but could well be connected with the method of pulse-height discrimination used.

7.2 THE EIGHT COMMON OXIDES

7.2.1 Silica (normally reported as SiO_2)

Silica occurs in most natural materials. It may be found from virtually the 100% level down to trace contents. In a 'full' analysis its determination is almost always called for.

$K\alpha$ line, InSb or PE crystal, coarse collimator, low kV/high mA, argon/methane flow proportional counter

The determination of silica is one of the most important in ceramic analysis; it is certainly one of the most common. The silicon $K\alpha$ line is measured using a coarse collimator and a flow counter. A PE crystal is the normal choice for sequential instruments, but for a fixed-channel instrument the indium antimonide crystal is far better. Not only is the latter crystal about four times as sensitive as PE, but it is less susceptible to temperature changes and has lower reflectivity for second order lines. Temperature stability is of great importance in this instance, as maximum accuracy is required in the determination of silica at high levels of content, and even minor variations of temperature could be significant. The detection limit with InSb is significantly better than for PE in spite of the fact that InSb suffers from a higher background.

When alumina is present in the sample as a major constituent, an α-correction has to be applied during calibration to allow for the effect of aluminium on silicon.

There is some line interference from the tungsten $M\alpha$ line which is relevant when grinding has been carried out in tungsten carbide. However, when using the Telsec TXRF instrument with an InSb crystal the overall effect was negative, owing to a reduction in the level of the background. Similar reductions in background caused by the presence of lead oxide and zirconia also gave rise to negative interferences.

7.2.2 Titania (normally reported as TiO_2)

Titania normally occurs naturally in the range of trace to minor contents except when dealing with titanium minerals such as rutile. It is found in a number of ceramic formulations such as glazes at between 'zero' and $\leqslant 10\%$ TiO_2 content and is an essential major content in technical materials such as barium titanates.

$K\alpha$ line, LiF 220 crystal, fine collimator, low kV/high mA, argon/methane flow proportional counter

Titania suffers from a serious line overlap from barium $L\alpha$. However, in most routine samples the level of BaO is low and the titanium $K\alpha$ line can be used with a LiF 200 crystal and fine collimation provided that a line overlap correction is made. Using a Telsec TXRF, 1% of BaO $\equiv 0.08\%$ of TiO_2; the effect on the goniometer of an ARL 8480 was found to be only 0.006% of TiO_2. Figures of a similar size have been found for the PW 1410 and the PW 1606. If high-level barium oxide contents occur, as in barytes, witherite or barium titanate, then the error becomes too great and other conditions must be used. The titanium $K\beta$ line suffers from even worse interference than the $K\alpha$ line (1% of BaO $\equiv 0.45\%$ of TiO_2 on the ARL 8480) due to the barium $L\beta_3$ line. The use of a LiF 220 crystal helps to overcome this problem (1% of BaO $\equiv 0.002\%$ of TiO_2 using this crystal and fine collimation on an ARL 8480). It is also possible to reduce the interference with a LiF 200 crystal on curved crystal optics if fine slits are used (1% of BaO $\equiv 0.003\%$ of TiO_2 on a PW 1606).

Hafnium $L\beta_1$, in the second order, also interferes with titanium $K\alpha$ (1% of $HfO_2 \equiv 0.01\%$ of TiO_2 on the Telsec TXRF). This must be corrected for when analysing zirconia (ores or refractories) and zircon-bearing materials. Scandium ($K\beta_{1,3}$) and lanthanum (a weak L_n) also have lines nearby, both causing only minor line overlap effects. There is also an iodine line (I $L\beta_{2,15}$) close by; this has to be borne in mind if lithium iodide or iodate is used as releasing agent.

7.2.3 Alumina (normally reported as Al_2O_3)

Like silica, alumina is found over the whole range of contents in natural materials, and may have to be determined in almost all samples at a variety of content levels.

ELEMENT LINE SELECTION

$K\alpha$ line, PE crystal, coarse collimator, low kV/high MA, argon/methane flow proportional counter

Aluminium, more than any other element, would benefit from a new crystal, as was the case for silicon before the appearance of InSb. Like silica, alumina is present in many ceramic and geological materials as a major component. Ideally, then, one is looking for high sensitivity so that good counting statistics can be achieved, yielding, in turn, good precision. PE, as already mentioned, suffers severely from instability to temperature changes. This demands constant vigilance with regard to ensuring constant temperatures in the spectrometer, particularly in the vicinity of the PE crystal. It was found advisable with the simultaneous Telsec TXRF to instal a temperature monitor in such a position with a digital read-out mounted prominently on the console. It is to be hoped that a solution to this problem may become available in the development of an appropriate layered crystal. Hopefully, the same solution would reduce the problems of interference from higher order lines, notably third-order barium $L\alpha$.

The above overlap from barium is serious in the analysis of materials such as barium carbonate, sulphate and titanate. The overlap is far worse with a chromium tube (1% of BaO $\equiv$ 0.09% of Al_2O_3, on the Philips PW 1410) than with a rhodium tube (1% of BaO $\equiv$ 0.007% of Al_2O_3, on the Telsec TXRF). Similarly, the interference using a scandium tube will be of the same order as that with a rhodium tube. There is, however, a scandium $K\beta_{1,3}$ third order peak nearby that could potentially give rise to background problems with a scandium tube, but the effect appears to be small. Another interference worthy of note is due to the bromine $L\alpha$, first order. This should give rise to problems only if a bromide salt is used as a 'releasing agent'. A fourth order chromium $K\beta_{1,3}$ line only 0.03° away could cause increased background if a chromium tube is used, but this is insignificant as a line interference from any Cr_2O_3 in the sample.

As with silicon, the presence of heavy elements depresses the background and produces a negative 'line interference'.

7.2.4 Ferric Oxide (normally reported as Fe_2O_3)

Iron oxides are almost universally present in natural materials in trace to minor content levels. Iron- and steel-making slags, together with iron ores and bauxites (raw), can contain major contents, as can some colours and brick-making clays.

$K\alpha$ line, LiF 200 crystal, fine collimator, high kV/low mA, krypton sealed counter

Settings for this element usually provide no problem. With Philips sequential instruments, flow and scintillation counters are usually used in tandem. On other instruments, where a single counter is used, a krypton sealed counter is preferable to an argon flow counter, as the awkward iron $K\alpha$/argon $K\alpha$ escape peak is eliminated and overall sensitivity is slightly improved. The most important line interference on iron is due to the manganese $K\beta_{1,3}$ line (1% of $Mn_3O_4 \equiv 0.008\%$ of Fe_2O_3, on the Telsec TXRF).

7.2.5 Lime (normally reported as CaO)

Lime is present in trace to minor contents in many natural materials, and is present at major content levels in limestones and dolomites. Many wall tiles will contain 5–10% of CaO, and most steel-making slags will have lime as one of their major constituents. It is also present as a major content in gypsum and its derivatives as well as in Portland cement. It is a main constituent in bone ash and will be present at around the 25% level in bone china.

$K\alpha$ line, LiF 200 crystal, coarse collimator, low kV/high mA, argon/methane flow proportional counter

Normally, coarse collimation with a flow counter is to be preferred for the calcium $K\alpha$ line. However, when high levels of lime are to be determined, e.g. in beads prepared from cement, dolomite or bone, at a flux/sample ratio of 5:1, the sensitivity of the calcium $K\alpha$ line can be too high when a chromium tube is used. As an alternative to turning down the power to avoid dead time and other counter problems, which is not always a realistic solution, it may be necessary to use fine collimation or even to resort to the calcium $K\beta_{1,3}$ line.

When analysing glazes, there can be problems from the tin $L\beta_3$ and $L\beta_4$ line (1% of $SnO_2 \equiv 0.025\%$ of CaO, on the Telsec TXRF). However, the maximum amount of tin that can be brought into solution in 4:1 flux is $\simeq 10\%$ of SnO_2 in the sample, so that the maximum background is equal to about 0.25% of CaO. The calcium $K\alpha$ line is very sensitive, so that fine collimation could be used in the presence of tin. There is also a cadmium $L\gamma$ line close by, but its effects should be considerably less than those due to tin.

7.2.6 Magnesia (normally reported as MgO)

Magnesia, like lime, is present in small amounts in many natural materials. It is present as the major constituent in magnesias and magnesites and, together with lime, is at major content level in dolomites and talc.

$K\alpha$ line, TlAP crystal or layered crystal, coarse collimation, low kV/high mA, argon/methane flow proportional counter

Thallium acid phthalate has now superseded ADP as the crystal of choice for magnesium, and could in its turn be replaced by a layered crystal. Although suffering less from high order interferences, the layered crystals are affected more by first order effects because of their poorer angular dispersion. For example, arsenic produces a 0.5% MgO interference for each 1% of As_2O_5 (measured on a PW 1606). This is four times the effect with TlAP. The chief problem with TlAP is line overlap from the nearby calcium $K\alpha$ third-order peak; a narrow pulse-height window may be used to reduce it. Using a chromium or a scandium tube the interference is far greater than with a rhodium tube. Therefore, if the chromium or scandium tubes are used then either a fine collimator should be employed or He/CO_2 should be substituted for Ar/CH_4 as the counter gas. As a comparison of the effects, a rhodium tube with coarse collimation gave an interference of 1% of $CaO \equiv 0.006\%$ of MgO on the Telsec TXRF, whereas a chromium tube with a fine collimator gave an interference of 1% of $CaO \equiv 0.002\%$ of MgO on the Philips PW 1410. There are also inconvenient high order barium lines nearby that create problems, particularly with a chromium tube, but with a rhodium tube the effects are quite small (1% of $BaO \equiv 0.002\%$ of MgO, on an ARL 8480).

Background effects are larger with a chromium tube and it is advisable to measure an off-peak background ($-1.25°\ 2\theta$ on TlAP). With a rhodium tube it is feasible to avoid the use of an off-peak background correction and compensate for the effects by applying 'line interference' corrections, either negative or positive. The worst effects are from the heavy elements, particularly zirconium, where the interference shows a quadratic relationship. Typical values measured on the Telsec TXRF are tabulated below (Table 7.2). Background and higher order reflections on MgO can be significantly reduced by applying a very tight pulse-height window setting. This may be so tight that it cuts out some of the tail of the magnesium $K\alpha$ peak itself.

7.2.7 Potash (normally reported as K_2O)

Potash is present at a minor content level in many common natural materials, and will be determinable at trace element level in many more. Fluxes such as feldspars may have contents up to about 15% of K_2O.

$K\alpha$ line, LiF 200 crystal, coarse collimator, low kV/high mA, argon/methane flow proportional counter

Table 7.2. Interferences on magnesia (using TlAP)

Interfering oxide (1%)	Effect (% MgO)
WO_3	-0.01
ZrO_2	$+0.02$
PbO	-0.006
ZnO	-0.005
As_2O_3	$+0.12$

This element poses very few problems. The only difficulty occurs when colours, glazes or enamels which contain cadmium are being analysed. The cadmium $L\beta_1$ line lies only 0.2° 2θ away. With coarse collimation the magnitude of the interference was found to be 1% of CdO $\equiv 0.13$% of K_2O, both for the Telsec TXRF and the ARL 8480. Use of a fine collimator will reduce the effect to about 0.10% but with the disadvantage of a sixfold loss in sensitivity. However, as this is a fairly sensitive line, the price may not be too high to pay for even the slightly reduced overlap. The potassium $K\beta$ line is of questionable value, as it suffers from an antimony $L\alpha$ line overlap, and antimony is occasionally found in colours, glazes and enamels. There is also a uranium $M\gamma$ and, in addition, a silver $L\gamma_1$ line close by that may give problems if the latter element is used as the material for the internal spectrometer mask.

7.2.8 Soda (normally reported as Na₂O)

The natural distribution of soda is very similar to that of potash, but at generally lower contents. It is more frequently used as an additive than potash because it is less expensive; hence its occurrence at higher levels than the latter in glazes and many glasses.

$K\alpha$ line, TlAP or layered crystal, coarse collimator, low kV/high mA, argon/methane flow proportional counter

Again, as for magnesia, the TlAP crystal is likely to be superseded by the new layered crystals. For the sodium $K\alpha$ line the gain in sensitivity will be even greater. High order reflections are again less of a problem. The use of a scandium tube will give a little more sensitivity. The same comments regarding background effects for chromium and rhodium tubes apply equally to sodium as magnesium (see magnesia, above). The background effects using a rhodium tube are higher than those on magnesium and extend to quite light elements (even potassium). However, even when no off-peak background can be measured, these effects can be corrected for. Therefore it is possible to achieve accurate results with a fixed-channel instrument where measurement of off-peak background is not feasible, provided that a rhodium tube is installed. As is the case with magnesia, great benefit is to be gained in terms of reduced background and high order interference by applying a very tight pulse-height window.

There are also some significant inter-elemental interferences on sodium, and these are tabulated below (Table 7.3).

The effect of zinc is worse on a layered crystal because of the lower 2θ value. On a PW 1606 it was found that 1% of ZnO $\equiv 0.4$% of Na_2O, three times greater than that with TlAP.

There are also second order phosphorus $K\alpha$ and $K\beta_1$ lines nearby, but these do not have much effect. In general, the sodium line is too insensitive to use with fine collimation in order to reduce these effects and they just have to be

ELEMENT LINE SELECTION

Table 7.3. Interferences on sodium $K\alpha$ line, using a TlAP crystal (measured on the ARL 8480)

Line	Order	Effect of 1% of element on Na_2O (TXRF)
Zinc $L\alpha$, $L\beta_1$	1st	0.14%
Zirconia $L\alpha_{1,2}$	2nd	0.06%
Barium $L\beta$	4th	0.008%
$M\gamma$	1st	
Strontium $L\beta_3$, $L\beta_4$	2nd	0.02%
Lanthanum $M\gamma$	1st	0.01%
Copper $L\beta_3$	1st	0.002%

suffered and corrected for. It may eventually prove possible with a scandium tube and a layered crystal to use fine collimation.

The line overlap due to zinc means that the determination of soda by XRF is very conditional on the requirements of the analysis. Where zinc is present in high minor or major contents the determination of sodium becomes very difficult. The levels of error introduced into the sodium determination cannot usually be regarded as acceptable. This is particularly true when, as is often the case, low levels of sodium are being sought. In fact, XRF is probably only of value for the determination of sodium at high levels of zinc content when the sole interest lies in finding out if abnormal levels of sodium are present when there should be very little. This means that recourse has to be made to conventional methods, such as flame photometry or ICP, for the determination of sodium in materials bearing a high level of zinc.

7.3 OTHER OXIDES/ELEMENTS

7.3.1 Fluorine (normally reported as F)

Fluorine is a common constituent in many ceramic materials, albeit at low levels of content. Most clays (ball clays, fireclays, etc.) contain up to about 0.05% of F. Fluxes such as feldspars may contain a slightly smaller amount, whereas Cornish stone has rather more, usually between 0.5 and 1.0% F. It is from this source that China clay derives contents of about 0.15–0.3% of F. Normal firing removes about one half of that present in bodies that do not contain raw materials having more than about 0.1% of F. Where large weights of ware are being fired, such as in the manufacture of bricks or tiles, the emission of fluorine, even at such apparently low levels, can lead to severe environmental problems.

As well as these normally occurring contents of fluorine, the element is deliberately added during the processing of a number of manufactured products. These include opal glass ($\simeq 5\%$), vitreous enamels and welding fluxes (5–15%). Calcium fluoride is commonly used in some of the processes for the manufacture of steel and is therefore found in a number of slags at moderate percentages.

$K\alpha$ line, TlAP crystal or layered pseudo-crystal, coarse collimation, low kV/high mA, flow proportional counter in the first order

Both because of its volatility during fusion, where it is easily lost as boron trifluoride (BF_3) and, in principle, as silicon tetrafluoride (SiF_4), and its lack of sensitivity on all but the most modern instruments, there is little experience concerning the actual determination of this element using XRF. With the current availability of instruments such as the ARL 8480 and the Philips 1606, both of which have reasonable sensitivity for fluorine, there is an incentive for attempting the determination. This means that samples containing fluorine have to be fused in such a manner as to attempt to retain all the fluorine, or, at least, cause a reproducibly minimal loss of fluorine. This is so that samples can be equated with standard beads fused in a similar manner. This requirement will restrict the fusion temperature to no greater than 1050°C and the method will be applicable only to suitable matrices.

The $K\alpha$ line is best used. The poor line to background ratio will dictate that at least one off-line peak background will need to be used. One appearing to be suitable is at $+4°$ 2θ.

Interferences so far noted are from zirconium ($L\alpha$, third order) and cobalt Ll.

7.3.2 Phosphorus Pentoxide (normally reported as P_2O_5)

Phosphorus is present as phosphate in many, if not most, ceramic materials as a trace or minor impurity, most frequently at levels about or below 0.1% P_2O_5. Additions of Calgon, a complex polyphosphate, are often made to pottery materials to help to deflocculate them during the making of slips. Phosphate is also a major constituent in bone and bone ash, and hence in bone china. In fact, the legal definition of the latter for export purposes to some countries requires the presence of a specified minimum amount of bone ash as ascertained from the phosphate content (P_2O_5). Some glasses and an increasing number of specialist products utilize phosphate in their formulation at minor or major content levels.

$K\alpha$ line, Ge crystal, coarse collimator, low kV/high mA, argon/methane flow proportional counter

A Ge crystal is used to give better dispersion and to eliminate the interference from the calcium $K\beta_1$ line in the second order. This is significant because, in many of the common ceramic materials where phosphorus pentoxide is found as a major content, e.g. bone ash and bone china, it is present together with a large amount of lime.

Notable interferences are due to appreciable levels of zirconium via its $L\alpha$ line (1% of $ZrO_2 \equiv 0.02\%$ of P_2O_5, on several spectrometers), molybdenum (Ll) and yttrium ($L\beta_1$), (again 1% of either oxide $\simeq 0.02\%$ of P_2O_5). Yttrium can be found associated with rare earths (see tungsten, chromium and manganese) but may also be a deliberate addition to a ceramic such as silicon nitride or stabilized zirconia. Molybdenum is normally only encountered in significant amounts in the analysis of Cermets (ceramic to metal seals). Scandium produces a very slight line overlap effect from its second order $K\alpha$ line, but this will only be significant if a PE crystal is used.

7.3.3 Sulphur Trioxide (normally reported as SO₃)

Sulphur is a very common element in ceramics, normally occurring either as sulphate or as sulphide. It is usually present in minor to trace element amounts in the generality of materials, but is met with in major contents in such materials as alkaline earth sulphates, e.g. gypsum, plaster and barytes, and in sulphide ores such as galena. As, at these higher levels of content, it is normally determined by methods other than XRF it is not necessary to discuss its determination in detail here.

Although it is only possible in certain circumstances to determine sulphur quantitatively by the fused, cast bead technique, there is nearly always sufficient left in the bead after fusion to indicate that an apparently low total can be attributed to the presence of this element. It is usually the most rewarding first check to make in such circumstances. Frequently, in materials where the sulphur is present as sulphate, an estimate of the sulphur in the bead produces a satisfactory analytical total and gives a reasonable indication of how much there was in the sample originally. This is particularly true when the situation is known before starting the analysis, so that attention can be given to using techniques designed to retain the sulphate in the bead. Furthermore, it is also feasible to determine sulphur (and for that matter, chlorine) using a pressed disc technique, provided that ultimate levels of accuracy are not required. This is particularly the case with cement raw meals and clinkers, where the figures achievable are satisfactory for the cement technologist.

$K\alpha$ line, Ge crystal, coarse collimator, low kV/high mA, argon/methane flow proportional counter

The considerations of the instrumental parameters to be used for sulphur are much the same as for lead $M\alpha$, p. 141, except that the sulphur $K\alpha$ line is more sensitive than the lead $M\alpha$ line. Here, however, there is no other line (such as the lead $L\beta$ line) to use as an alternative. Again the Ge crystal is to be preferred to the PE crystal because of its better resolution and temperature stability. There is again a slight penalty to pay in sensitivity, but this only amounts to about an 8% drop. The power settings to be used are low kV/high mA together with a flow counter. The important interferents and their relative effects on Ge and PE crystals are tabulated below (Table 7.4).

It is possible, using fine collimation, to reduce these interferences to insignificant levels, even that from lead, but the four-fold loss in sensitivity is usually too high a price to pay. For ceramic analysis it has not been found necessary to attempt to determine sulphur in the presence of high levels of lead and/or molybdenum. Therefore coarse collimation has always been used.

The effect of the cobalt $K\alpha$ third order line on the sulphur $K\alpha$ line is similar to the effect of the iron $K\beta_{1,3}$ on lead $M\alpha$, and can be reduced in a similar way by applying a tight pulse-height window. Being an α line means that its effect is even greater, and merely reducing the window from 3 to 2 on a threshold of 5 will only reduce the interference using a Ge crystal by about a quarter. A window of 1 on a threshold of 6 drops the interference to a quarter, but so tight a window is not very practical because of the problems of setting the EHT on the counter sufficiently accurately. Fortunately, in practice, it has not been found necessary to determine sulphur in the presence of large levels of cobalt, nor is this combination likely to occur when the fused, cast bead technique is being used. If it were necessary to do this then the PE crystal would be preferable. Note again that, although the Ge crystal eliminates second order (and other even order) interferences, those for third order are worse than for other crystals.

7.3.4 Chlorine (normally reported as Cl)

Traces of chlorine are widespread at low levels, in an analogous manner to fluorine. In most instances, the presence of chlorine at these low levels is of less consequence than that of fluorine, so that its determination is less frequently required. It is, however, present in significant amounts in clays from some areas,

Table 7.4. Interferences on sulphur $K\alpha$ line (effect of 1% interferent, measured on an ARL 8480)

Interferent	Line	Crystal	
		Ge	PE
PbO	$M\alpha$	0.013	0.072
MoO_3	$L\alpha$	0.195	0.327
ZrO_2	$L\gamma_1$	0.024	0.022
Co_3O_4	$K\alpha$ 3rd	0.052[a]	0.004[a]
WO_3	M_2-N_4	0.004	0.006

[a] These interferences are ten times smaller on the TXRF and the PW 1606.

notably the Middle East, where the presence of sodium or other chlorides is not uncommon at levels liable to give rise to difficulties during 'wet' processing and firing of the material. Contents can reach into the percentage range, at which level problems may arise with deflocculation of slips and again later during the firing process, when chlorine and hydrochloric acid are emitted. These gases cause environmental difficulties and have been responsible for rapid deterioration of the kiln itself by attack on the metalwork, etc., particularly iron and steel components.

$K\alpha$ line, Ge crystal, coarse collimation, low kV/high mA, flow proportional counter

It is possible to use either PE or Ge crystals for the determination of chlorine. Although not the more sensitive crystal, Ge has the edge over PE, as it has a better resolving power as well as significantly better temperature stability and lack of second order interferences. Ge is particularly advantageous, for the latter reason, when a chromium target tube is used because of the presence of the second order chromium $K\alpha$ peak with the PE crystal.

A problem specific to chlorine is that the element is easily picked up by the metal surface of the masks and other internal metal parts of the spectrometer. This increases the apparent measured background, thereby increasing detection limits. Thus, if low levels of chlorine are to be determined, great care should be taken to clean all such surfaces. Copper masks appear to be more prone to this contamination than those made of aluminium.

The use of a Ge crystal eliminates second order interferences from neodymium $L\alpha$, and the third order hafnium $L\alpha$ also causes no interference despite the crystal's greater reflectivity in the third order. There is also a slight interference from some weak third order bismuth lines (1% of $Bi_2O_3 \equiv 0.006\%$ of Cl).

7.3.5 Scandium Oxide (normally reported as Sc_2O_3)

Scandium is not normally found in ceramics, except when deliberately added, and even this is infrequent. Thus, normally it will be present only in the trace element range and, even so, in the lower ranges, and will thus demand special trace element techniques for measurement.

$K\alpha$ line, LiF 200 crystal, fine collimation, low kV/high mA, flow proportional counter

There do not seem to be any special problems with this element unless, of course, a tube with a scandium target is used, when a filter is necessary.

Antimony interferes via the $L\beta_{2,15}$ line so that 1% of $Sb_2O_3 \equiv 0.008\%$ of Sc_2O_3.

It should be noted that scandium itself interferes with barium via its $K\beta_{1,3}$ line (1% of $Sc_2O_3 \equiv 0.02\%$ of BaO).

7.3.6 Vanadium Pentoxide (normally reported as V_2O_5)

Vanadium pentoxide is not infrequently found in clays at or below the 0.1% level. Similar levels may be found in fired materials produced in kilns fired by fuel oil. It can give rise to soluble salt problems, particularly if it is in the form of vanadate, as it usually is after firing. This is almost always so when the body contains significant amounts of basic material, e.g. lime, as in wall tiles. The soluble vanadium (usually as vanadate) tends to migrate on drying, either during processing or in service, giving unacceptable areas of yellow or orange staining. In addition, it may be present as a deliberate addition in glazes, stains or colours, the processing of which demands great care to prevent the formation, or removal by washing, of soluble vanadates.

$K\alpha$ line, LiF 220 crystal, fine collimation, high kV/low mA, krypton sealed counter

The main problem with this element arises from the first order interference due to the overlap of titanium $K\beta_{1,3}$. This is best handled by using a LiF 220 crystal; the magnitude of the overlap is then 1% of $TiO_2 \equiv 0.01\%$ of V_2O_5. Even at this level the interference is significant in practical terms, as the level of vanadium pentoxide is often well below 0.1% and is found in the presence of 1-3% of TiO_2. If a LiF 220 crystal is not available, it has been found that a PE crystal used in the second order can reduce the overlap, but at the expense of some sensitivity. On a Philips 1410 the interference from TiO_2 was reduced from the equivalent of 0.11% of V_2O_5 to 0.04% of V_2O_5 using this approach.

A krypton counter will give a better sensitivity than a flow counter, albeit with a theoretical risk of third order reflections, none of which have been encountered in practice. Vanadium $K\alpha$ occupies a useful position where the kV setting, at least in the range 30-60 kV, is not critical.

There is also an interference by vanadium $K\beta_{1,3}$ on chromium $K\alpha$ that can be significant in the determination of the latter (1% of $V_2O_5 \equiv 0.05\%$ of Cr_2O_3, on the Telsec TXRF).

7.3.7 Chromium Sesquioxide (normally reported as Cr_2O_3)

Chromium sesquioxide is a common constituent of many ceramic materials, occurring naturally in amounts ranging from traces up to 0.1% content. The main natural exception to this is chrome ore, which is used as a raw material, mainly for basic refractories. It is added to magnesites in amounts sufficient to produce chromium sesquioxide contents ranging from about 5-40%. The materials concerned, in increasing order of Cr_2O_3 content, are magnesites, magnesite-chrome and chrome-magnesite refractories. Chromium sesquioxide, added either as the ore or as a prepared 'pure' oxide, may also be found in coloured glasses and glazes, vitreous enamels, colours and stains. Slags may

ELEMENT LINE SELECTION

contain either small or significant amounts depending on their source. The determination of chromium in glass sands is very important, but as the content generally needs to be less than 5 ppm it will normally require other techniques.

$K\alpha$ line, LiF 200 crystal, fine collimator, high kV/low mA, krypton sealed counter

The XRF conditions for chromium are straightforward, the $K\alpha$ line and either a flow proportional counter or a krypton sealed counter can be used together with fine collimation and a LiF 200 crystal. Obviously, if a chromium tube is used, a filter, usually aluminium foil, is required. Therefore, rhodium or scandium tubes are preferable if chromium is to be determined without serious loss in sensitivity. If a choice of detectors is available, krypton will be preferable to an argon flow proportional counter because the peak and escape peak are combined for krypton and hence better sensitivity is achieved. Unlike manganese, chromium does not suffer any serious third order overlaps that can slip into the pulse-height window. The most common interference is that due to vanadium $K\beta_{1,3}$, first order, but fortunately vanadium is rarely met with in amounts much in excess of 0.5% in ceramic materials. On a Telsec TXRF the interference was found to be 1% of $V_2O_5 \equiv 0.05\%$ of Cr_2O_3, and this sort of effect is easily corrected for mathematically. Both the rare earths cerium $L\beta_3$ (1% of $CeO_2 \equiv 0.003\%$ of Cr_2O_3, on an ARL 8480) and lanthanum $L\beta_{2,15}$ (1% of $La_2O_3 \equiv 0.009\%$ of Cr_2O_3, on an ARL 8480; 1% of $La_2O_3 \equiv 0.01\%$ of Cr_2O_3, on a PW 1606) interfere with chromium. As discussed in the next section, rare earths also interfere with manganese. Therefore, if it appears that small amounts of chromium or manganese are present in a sample together with a low analytical total, and without the bead showing the characteristic green or brown coloration normally caused by the presence of these elements, this is a very strong indication of the presence of rare earths.

7.3.8 Manganic Oxide (normally reported as Mn_3O_4)

Manganese occurs in small amounts (less than 0.2%) in a very wide range of materials, particularly fireclays and some types of brick-making clay. It is also widely used in glasses, glazes, vitreous enamels, colours and stains. Slags may contain any amount from trace levels to several per cent. Manganese dioxide is the most frequently used additive either in the natural state (as the ore, pyrolusite) or after treatment, and the assay of this additive is occasionally required.

$K\alpha$ line, LiF 200 crystal, fine collimator, high kV/low mA, krypton sealed counter

As with chromium, a fine collimator is used with a LiF 200 crystal and the $K\alpha$ line. Either a flow counter or a flow plus scintillation counter can be used.

Although more sensitive than a flow counter, a krypton sealed counter is less satisfactory because of the presence of third order zirconium $K\beta_1$ and $K\beta_3$ lines only 0.07° and 0.13° 2θ away. Therefore, if high levels of zirconia are present the krypton counter cannot be used. Using the ARL 8480, the line overlap due to zirconia on a krypton counter is in the order of 0.02% per 1% of ZrO_2, but with a flow proportional counter it is only about 0.0002% per 1% of ZrO_2. Similar problems of tube line interference arise with the chromium target tube as are experienced with chromium itself, and therefore this tube is better avoided if possible. Alternatively, a primary filter may be used. The most significant interference is due to chromium $K\beta_1$ and was found to be 0.02% per 1% of Cr_2O_3 on a Telsec TXRF instrument. For most materials a mathematical correction is sufficient, but in the presence of high levels of chrome the LiF 220 crystal must be used. Manganese also suffers from interference from neodymium $L\beta_6$ (1% of $Nd_2O_3 \equiv 0.005\%$ of Mn_2O_3, on an ARL 8480) and $L\beta_3$ and praseodymium $L\beta_{2,15}$ (1% of $Pr_2O_3 \equiv 0.005\%$ of Mn_2O_3, on an ARL 8480). As discussed in the previous section, an unexpected apparent presence of manganese together with a low analytical total could indicate the presence of rare earths in the sample.

7.3.9 Cobalt Oxide (normally reported as CoO)

Cobalt in anything but lower trace element levels is unusual in most naturally occurring materials. However, it is frequently met as a deliberate addition, and it will usually be found in samples if the material has been ground in a tungsten carbide mortar, where it is used as a binder. In the latter case it will normally be at low and probably insignificant levels unless a considerable amount of contamination has been introduced during the grinding. Cobalt may be used as an additive in bodies or glazes in relatively small amounts to neutralize the colour caused by the presence of ferric oxide in the raw materials, or in larger amounts to produce a blue colour in a glass, glaze etc. It may also be mixed with other oxides to increase the extinction of light in the formulation of a black colour, again in a glassy matrix. The oxide is, therefore, often found in coloured glasses, glazes, colours and vitreous enamels, but its use in glaze and colours is less frequent nowadays as it reduces the resistance of the glaze to acid attack, making it liable to release greater quantities of lead.

It should be noted that if cobalt is called for in materials of the first category or at trace element level, the use of tungsten carbide needs to be avoided unless a nickel bonded vial is used.

$K\alpha$ line, LiF 200 crystal, fine collimation, high kV/low mA, krypton sealed counter

A krypton counter would be used on its own, but flow and scintillation counters are suitable if the two can be used in tandem.

No serious interferences have been noted, but iron $K\beta_{1,3}$ introduces a small interference (1% of $Fe_2O_3 \equiv 0.013\%$ of CoO); this may need to be taken account of with materials containing smaller levels of cobalt oxide or in the analysis of iron oxide. There is also an interference due to hafnium Ll (1% of $HfO_2 \equiv 0.002\%$ of CoO); this is rarely of importance since, even when hafnium is present, no more than a 2% content is to be expected.

7.3.10 Nickel Oxide (normally reported as NiO)

Nickel is rarely found naturally in ceramic materials except in trace amounts, but may occur at low levels in materials that have been fired using fuel oil. It is used as a colouring agent to produce a green colour in a glassy matrix, and together with other oxides to produce a black colour; cf. cobalt. It may well, therefore, be found in coloured glazes, glasses, colours and vitreous enamels.

$K\alpha$ line, LiF 200 crystal, fine collimation, high kV/low mA, krypton sealed counter

A krypton detector is normally preferable to a flow proportional counter because it yields better sensitivity.

There are two second order overlaps that cause problems. The first is due to rubidium $K\beta_{1,3}$ (1% of $Rb_2O \equiv 0.02\%$ of NiO) and the other to yttrium $K\alpha$ (1% of $Y_2O_3 \equiv 0.1\%$ of NiO). Neither of these elements is commonly present in significant amounts and would only affect the determination of nickel at trace contents; where nickel is being determined after deliberate addition, it is unlikely to be affected by the small contents of these two interferents. On the few occasions where account has to be taken of them it is necessary to revert to the use of an argon/methane flow proportional counter, which greatly reduces the second order overlaps in this region of the spectrum. By this means the two interferences are reduced to 1% of $Rb_2O \equiv 0.003\%$ of NiO and 1% of $Y_2O_3 = 0.03\%$ of NiO.

One element that interferes in the first order is ytterbium, through the ytterbium $L\alpha$ line, which is about $+0.4°$ 2θ away on a LiF 200 crystal. The extent of the effect is 1% of $Yb_2O_3 \equiv 0.006\%$ of NiO; this can be reduced, at the expense of some reduction in sensitivity, by using a LiF 220 crystal.

An interference that might be expected is from the second order lead $L\gamma_2$, but this has not been encountered in practice. The obvious first order interference from cobalt $K\beta_{1,3}$ is quite small (1% of $Co_3O_4 \equiv 0.001\%$ of NiO).

Some technical ceramics, particularly zirconias, have additions of either yttrium or ytterbium oxide. If the analyst is unaware of the problem he could report the presence of nickel when it is not, in fact, present. If, in these instances, a low analytical total is obtained, together with the apparent presence of nickel (but without a typical greenish coloration of the bead), yttrium and/or ytterbium should be sought.

7.3.11 Cupric Oxide (normally reported as CuO)

Copper oxide is normally found in natural ceramic materials at the trace level. A few fireclays may contain chalcopyrite (copper pyrites) in the form of nodules. These cause the fault of 'green spot' in sanitary ware, causing an effect that may be deduced from its name, i.e. the formation of a spot in the glaze which is often black with a green halo or sometimes merely a green spot or circle. Even when this fault is found, the average level of content of copper oxide in the material as a whole is still very low. Apart from this, copper is often added as a deliberate component to impart a green colour to a glassy matrix or to help form a black hue; cf. cobalt. Its use has diminished in recent years, since its presence enhances the danger of lead release from the glaze or colour. It is found in some slags in significant amounts and also in superconducting materials.

$K\alpha$ line, LiF 200 crystal, fine collimation, high kV/low mA, krypton sealed counter

Provided that the copper oxide content of the sample does not exceed about 3%, little difficulty should be encountered during fusion, although a releasing agent such as lithium iodide or iodate may be needed to obtain a clean release of the bead from the casting mould. This content is about the limit that can be handled by the fused, cast bead technique because of the tendency of the bead to stick (and often crack), so that if this content is exceeded it will amost certainly prove necessary to carry out dilutions.

Like nickel, some higher order reflections may cause problems if a krypton counter is used. However, again like nickel, one would have to encounter a peculiar combination of elements for the problems to arise. The interferences concerned are those from strontium $K\beta_2$ and strontium $K\beta_{1,3}$ in the second order, (1% of SrO $\equiv$ 0.001% of CuO) and zirconium $K\alpha$, also in the second order (1% of $ZrO_2 \equiv$ 0.0015% of CuO).

Another less important interference may be mentioned. With modern industrial ceramics, a first order reflection may be noted from tantalum $L\alpha$ (1% of $Ta_2O_3 \equiv$ 0.005% of CuO).

Finally, an interference caused by nickel, which may well be met with in analysing highly coloured samples or black colours in a glassy matrix, is due to the nickel $K\beta_{1,3}$ line, where 1% of NiO $\equiv$ 0.001% of CuO.

7.3.12 Zinc Oxide (normally reported as ZnO)

This element is found, except in minor trace element amounts, chiefly in glazes, glasses, colours, ferrites and some slags. As the trace element levels are below commercially practical detection limits they are of no concern here.

$K\alpha$ line, LiF 200 crystal, fine collimation, high kV/low mA, krypton sealed counter

The conditions for the determination of zinc are straightforward, i.e. high kV (60) and low mA (40–50), with fine collimation and either a scintillation counter, flow plus scintillation counter (Philips PW 1400 series) or, on fixed-channel instruments, a xenon proportional counter. Although the $K\alpha$ line is in a region of the spectrum fairly thick in lines, it is relatively free from interference, and a LiF 200 crystal can be used in preference to the LiF 220 to give better sensitivity. There is a gold Ll line nearby, but this should not cause much trouble unless a gold mask is used which is in the X-ray beam. In this case the LiF 220 crystal may be necessary in order to reduce background. There are two interferences that may be encountered, one from tungsten $L\alpha$ and the other from second order zirconium $K\beta_1$. Both these interferences are very small, as is that from Yb $L\beta_3$. The former is of consequence only if the sample contains $>3-4\%$ of WO_3 together with poorish dispersion (e.g. Telsec TXRF, fixed-channel). Therefore in practice it can be ignored. The effect of zirconia is only significant when analysing high-zirconia content materials, e.g. zircons, where 65% of ZrO_2 will show up as about 0.04% of ZnO.

One interference likely to be found only when analysing high-molybdenum content samples is that due to the second order molybdenum $K\alpha$ peak. In practice, MoO_3 is either found at very low levels (if at all) or is at very high levels, as in Cermets. In the unlikely event of zinc needing to be determined in such samples, then steps can be taken to reduce the effect (1% of $MoO_3 \equiv 0.01\%$ of ZnO, using LiF 200 and a scintillation counter). The LiF 220 crystal will reduce this effect (from 0.013% to 0.002% per 1% of MoO_3 in the case of the ARL 8480).

7.3.13 Gallium Oxide (normally reported as Ga_2O_3)

Gallium is almost never found as a deliberate addition, so that it is only met with at the trace or lower minor content level, usually in samples of high alumina content. It appears to associate in nature with alumina. Some gallium is thus normally found in high-alumina content materials, e.g. bauxites. It appears at present to be of little ceramic interest, and its determination is therefore rarely called for, but it could well be of geochemical interest.

$K\alpha$ line, LiF 200 crystal, fine collimation, high kV/low mA, scintillation counter

In this general area of the spectrum, there are a number of interferences which may prove significant, as the level of content sought is very low. The interferences shown in Table 7.5 have been noted using a scintillation counter.

As gallium oxide is only present at trace levels (typically 150 ppm in alumina), it is necessary to measure an off-peak background. A suitable position is at $37.60°$ 2θ if a LiF 200 crystal is used. Because there is some second order bremsstrahlung at this angle, it is advisable to subtract only a proportion of the background. For this reason, on modern spectrometers the background

Table 7.5. Significant interferences on gallium

Element	Line	Order	Interference (1% of oxide ≡ % Ga_2O_3)
Hf	$L\beta_2$	1	0.006
Ta	$L\beta_1$	1	0.01
Pb	Ll	1	0.0007
Nb	$K\beta_{1,3}$	2	0.003

should not be set up in the line parameters, but as a line in its own right. The relevant proportion of the intensity is then subtracted as an intensity overlap.

7.3.14 Germanium Oxide (normally reported as GeO_2)

Germanium rarely occurs in ordinary ceramic samples, even as an addition, but is often found in coal ashes or fly ashes and thus may occur in kiln or similar deposits etc., or even in bricks or blocks to which an addition of PFA (pulverised fuel ash) has been made. It is also widely encountered in the analysis of Cermets (materials for bonding ceramic to metal).

$K\alpha$ line, LiF 200 crystal, fine collimation, high kV/low mA, scintillation counter

Germanium suffers from special background problems due to the presence of the nearby second order rhodium $K\alpha$ line when a rhodium target tube is used. There are a few other interferences that have been encountered. The first is due to the molybdenum $K\beta_2$ and $K\beta_{1,3}$ in the second order, 1% of $MoO_3 \equiv 0.002\%$ of GeO_2. In addition, there is interference from tungsten $L\beta_2$ and $L\beta3$ in the first order, 1% of $WO_3 \equiv 0.07\%$ of GeO_2. The former can be a problem in the analysis of Cermets, the latter if grinding contamination is high. Where interference presents a problem, a LiF 220 crystal should be used.

There is also a nearby gold line, $L\alpha$, that could give rise to problems if a gold mask is in use.

7.3.15 Arsenious Oxide (normally reported as As_2O_3)

Arsenic occurs naturally in ceramic materials only in trace amounts. It is present in small amounts in coal and therefore tends to accumulate in the ash, or more especially at points in the kiln or exhaust system where the temperature has diminished sufficiently for volatile products to condense. It will be found in some glasses (e.g. lead crystal), where it is used as a decolorizer, usually at a level of about 0.5% of As_2O_3, and it may occasionally be found in colours, particularly yellows, where the additive is Naples yellow, lead arsenate. This latter use is much rarer of recent years for safety reasons and with the ready availability of good alternatives.

ELEMENT LINE SELECTION

$K\beta$ line, LiF 200 crystal, fine collimation, high kV/low mA, scintillation counter

The $K\beta$ line is generally used for this element, as the $K\alpha$ is seriously interfered with by a lead $L\alpha$ line; there is, of course a loss in sensitivity, but as the two elements commonly occur together there is little practical alternative. Lines both from a rhodium target (second order) and from a gold mask could give rise to problems of background in this region of the spectrum. Interference from tungsten $L\alpha$ is in practice insignificant, but that from cadmium $K\alpha$ in the second order can cause noticeable interference; 1% of $CdO \equiv 0.003\%$ of As_2O_3.

If interferences are a problem, LiF 220 may be used, but with a three-fold loss in sensitivity.

7.3.16 Bromine (normally reported as Br)

Bromine would normally only be expected in ceramic materials at trace levels, and these levels should be even lower after firing. It has been detected in recent years in sanitary ware kiln deposits because of the increasing use of plastic to assist setting in the kiln. The material used tends to contain bromine, which appears to be introduced as part of the plasticizer.

$K\alpha$ line, LiF 200 crystal, fine collimation, high kV/low mA, scintillation counter

This element, which is rarely called for and, in any event, is highly volatile, has not been determined in samples by XRF using the fused, cast bead method. Its determination is only attempted when it has been used in the bead releasing agent. When samples have been prepared as pressed discs the $K\alpha$ line is preferred, combined with the use of a LiF 200 crystal, fine collimation and a scintillation counter. If the sample is in the form of a thin layer, for example spread on to a filter paper, the $L\alpha$ line with flow counter, coarse collimation and a PE crystal is better unless the sample contains much alumina or a tube with a chrome target is used. Aluminium has a nearby line in the first order and chromium one in the fourth.

7.3.17 Rubidia (normally reported as Rb_2O)

The element is infrequently found in ceramic materials at more than trace levels, possibly rising to lower minor content levels in one or two fluxes such as feldspars.

$K\alpha$ line, LiF 200 crystal, fine collimation, high kV/low mA, scintillation counter

Those interferences that have any significance are shown in Table 7.6.

There are a number of other interferences but none of these have proved to be significant. They include thorium $L\alpha$ in the first order, antimony $L\alpha$, Pb $L\beta_5$ and thorium $L\alpha$ in the second order, and praseodymium $K\beta$, cerium $K\beta$ and lanthanum $K\beta_2$ in the third order.

Table 7.6. Significant interferences on rubidium

Element	Order	Line	Magnitude of Interference (1% of oxide ≡ % Rb$_2$O)
U	1	$L\alpha$	0.01
Bi	1	$L\beta_3, L\beta_5$	0.004
Cd	2	$K\beta_2$	0.0004

If high levels of rubidia need to be determined using the $L\alpha$ line, an InSb crystal, if available, will yield sensitivities about three times greater than a PE crystal as well as having the added advantage of greater temperature stability.

7.3.18 Strontia (normally reported as SrO)

Strontia in significant amounts is a less commonly occurring constituent of ceramic materials than the other alkaline earths. However, its determination may be required at trace or minor contents in a wide variety of materials. It will certainly be required in calcium or barium-bearing minerals, where contents up to and around 1% may be present. The same is true in the analysis of some feldspars, where sufficient SrO may be present to affect the analytical total. It is added in its own right, albeit infrequently, to glazes, glasses etc. Finally, it will be present in major contents in strontium minerals such as strontianite and celestine.

Kα line, LiF 200 crystal, fine collimator, high kV/low mA, scintillation counter

For high levels of content the $L\alpha$ line is preferable, for the same reasons as lead $M\alpha$ is preferable in the case of high levels of lead oxide and zirconium $L\alpha$ for zirconia. For most analyses the determination of SrO is not required at less than about 0.01%, so the $L\alpha$ line can be used throughout provided the spectrometer gives sufficient resolution from the silicon $K\beta$ line.

7.3.18.1 Strontium Kα

The conditions are straightforward, high kV/low mA, fine collimation using a LiF 200 crystal and a scintillation counter. At low levels an off-peak background (at $+1.04°$ 2θ) may be necessary.

7.3.18.2 Strontium Lα

This line is dominated by the effects of the nearby silicon $K\beta_1$ line with its Si K satellite and, as SiO$_2$ is nearly always a major constituent in ceramic materials, this will almost certainly produce a background effect. The conditions for the line are low kV/high mA, a flow detector and a PE crystal with fine

ELEMENT LINE SELECTION 127

collimation. The silicon peak occurs at $-2.5°$ 2θ distance. Even with fine collimation a spectrometer with good resolution is required. This was not the case with a goniometer-type Telsec TXRF but was with an ARL 8480. For the latter the effect of silica was found to be that 1% of SiO_2 gave an interference equivalent to 0.002% of SrO. This means that 100% of SiO_2 in a silica material has a background equivalent to 0.2% of SrO. In practice this sort of effect can be handled. An apparently obvious way of compensating for its effect would be to set up an off-peak background at the opposite side of the silicon peak to the strontium peak and at an equal background intensity. Unfortunately, at this position there is the second order calcium $K\alpha$ peak. Thus, the only solution is to calibrate SrO in an alumina matrix and apply a line overlap correction for the effect of SiO_2 and also its α-correction. For most analysts, who are unlikely to need to determine SrO above 1%, the use of the $K\alpha$ line will involve less effort.

The use of an InSb crystal for the $L\alpha$ line, if available, will increase the sensitivity by a factor of almost two over PE. The silicon line overlap is slightly reduced (1% of $SiO_2 \equiv 0.0016\%$ of SrO, on a PW1606) and, of course, there is the advantage of greater temperature stability of the crystal.

7.3.19. Yttrium Oxide (normally reported as Y_2O_3)

Yttrium occurs naturally in ceramic materials only rarely, except at low trace levels. The exceptions are in conjunction with high levels of zirconia in poor quality raw materials used as a source of the latter. It is also used as a deliberate addition at minor constituent levels, for example, in stabilized zirconias and in some magnesias used as abrasives.

$K\alpha$ line, LiF 200 crystal, fine collimator, high kV/low mA, scintillation counter

For low levels the $K\alpha$ line is used with normal parameters. Where yttria may be expected to be present at minor content levels the $L\alpha$ line is to be preferred. As this is the usual situation in which the determination of yttria is requested, most experience is with the $L\alpha$ line. The wavelength is outside the upper limit of the Ge crystal, so the less satisfactory PE crystal has to be used with most sequential spectrometers. The InSb crystal is preferable to PE but is rarely available on sequential spectrometers, as its use normally tends to be restricted to the silicon $K\alpha$, strontium $L\alpha$ and yttrium $L\alpha$ lines only. Coarse collimation is used for either crystal, together with a flow counter.

Some interferences on yttrium $L\alpha$ have been investigated using an ARL 8480 instrument but with a PE crystal. A first order interference has been noted from strontium $L\beta_3$, $L\beta_4$ and $L\beta_6$ (1% of SrO $\equiv 0.03\%$ of Y_2O_3) and a second order one from antimony $L\beta_1$ (1% of $Sb_2O_3 \equiv 0.0015\%$ of Y_2O_3). No significant effects were found from tin $L\beta_{2,15}$ in the second order or from praseodymium

$L\beta_{2,15}$, neodymium $L\beta_{1,4}$, barium $L\gamma_2$ and $L\gamma_3$ or lanthanum $L\gamma_1$ in the third order. Tungsten interferes with the yttrium $L\alpha$ (1% of $WO_3 \equiv 0.002\%$ of Y_2O_3) via its $M\beta$ line in the first order which, although 6° away, is a strong line and less well resolved with the coarse collimator used.

7.3.20 Zirconia (normally reported as ZrO_2)

In the field of ceramics, it would be unusual to need to determine this oxide at levels much less than 0.05% except in the glass industry, where this level could be of interest as a contaminant arising from the use of zircon-bearing refractories in the furnaces. On the other hand, it is often determined as a major constituent in zirconia ores and refractories, zircon and zircon-bearing refractories. At lower levels of content it is used in kiln furniture for the pottery industry, and in glazes, glasses and vitreous enamels as an opacifier. In glazes it is usually used in combination with zinc as an effective but cheaper replacement for tin. Therefore calibrations for this oxide need to be made for both major and minor contents. It is worth bearing in mind that when zirconia is present at minor or major content levels hafnia will also need to be determined.

$L\alpha$ line, Ge crystal, coarse collimator, low kV/high mA, flow counter

The advantages of long wavelength lines over short wavelength ones have been discussed. As this is the first important oxide of this type to be discussed, this point needs to be covered in more detail. Similar comments will be applicable in various degrees to other oxides discussed in this chapter, e.g. PbO, SrO, and MoO_3. The calibration for zirconia in the range 0–65% for zircon, using the $K\alpha$ line, will be so curved as to be unusable. It is also possible that the points on such a calibration curve will show scatter. One reason for the high curvature is count loss due to dead time and other effects. It could be argued that these can be corrected for, but this is a dangerous concept: although the corrected graph may be of low curvature, the actual data being used is, in fact, showing a poor response at the top end, and the analytical results will be due more to mathematical manipulation than to measured intensity. Even if the power is reduced (say, from a typical 60 kV/50 mA) to overcome effects in the counter, the calibration is still markedly curved. The actual sensitivity at the top end as counts/s/% is lower with the $K\alpha$ line than with the $L\alpha$ line. The detection limit of the K line is only about half that of the L line. Detection limits measured on an ARL 8480 (using a 10 s count) were 0.0026% for the $K\alpha$ line at 45 kV/20 mA and 0.006% for the $L\alpha$ at 30 kV/100 mA. The calibration using the $L\alpha$ line was, furthermore, a straight line from 0–65% of ZrO_2.

A further advantage of the zirconium $L\alpha$ line over the $K\alpha$ is that α-corrections are much lower on the former. The difference can be an order of magnitude or more. This means that the corrections do not have to be determined to the same degree of accuracy and in some cases need not be determined at all. Because the background is generally lower for $L\alpha$, background effects are smaller and the need for their correction is less.

It is worthwhile considering in some detail the actual magnitude of these large α-corrections for the zirconium $K\alpha$ and those of similar wavelengths. This is best illustrated by the case of the effect of lead on zirconia.

If a zircon sand of ZrO_2 content = 65.00% were to contain only 0.1% of PbO, the apparent ZrO_2 content would be 64.74%; an error of 0.26%. This could occur when the lead oxide was not being measured and would thus be uncorrected for. The PbO is causing an error of two and a half times its own concentration, an intolerable situation. If the $L\alpha$ line is used the error would only be 0.01% of ZrO_2. In a glaze-type matrix it would not be inconceivable to have 10.00% of ZrO_2 in the presence of 20% of PbO. The apparent zirconia content measured (without correction) on the $K\alpha$ line would be 3.46% ZrO_2 while that for the $L\alpha$ line would be 9.72%. Thus, if the $K\alpha$ line is used the correction for the lead oxide content accounts for twice as much of the analyte determined as that analyte's measured intensity; this situation is clearly not acceptable.

Having chosen the $L\alpha$ line as the most suitable one, the next consideration is to choose the optimum analytical parameters. The chief problem with this line is the overlap from phosphorus $K\alpha$. The Ge crystal is preferable to the PE crystal here because of its better angular dispersion and, hence, higher resolution. Crystal fluorescence is not a problem in the wavelength range for which it is used because the germanium/argon escape peaks are safely above the pulse-height window that would normally be set. The weaker germanium $L\alpha$ line would be in, or close to, the background noise cut out by the threshold setting. A coarse collimator would normally be used, as the four-fold loss in intensity caused by using a fine collimator is unacceptable. Using a Ge crystal rather than PE reduces the line overlap effect for 1% of P_2O_5 from 0.23% to 0.02% and nearly doubles the sensitivity. On a PW 1606 with curved crystal optics the effect was reduced to 0.003%. The level of interference of phosphorus on zirconium using a Ge crystal and coarse collimation is normally acceptable because, in the authors' experience, lower-level zirconia contents are not normally required when the phosphorus pentoxide contents are high. In practice the line overlap of strontium $K\beta$ on zirconium $K\alpha$ (1% of SrO $\equiv$ 0.08% of ZrO_2) would be more of a problem. There is a second order scandium $K\alpha$ peak quite close to the zirconium $K\alpha$. This, of course, does not cause a problem with a scandium tube if a Ge crystal is used, but could be serious if a PE crystal were chosen.

7.3.21 Niobium Pentoxide (normally reported as Nb_2O_5)

Niobium will normally occur only at very low trace levels. Its use as an additive is almost unknown in normal ceramics, but it is feasible that it could occur as a contaminant after service and it is therefore a possible request albeit very infrequently.

$K\alpha$ line, LiF 200 crystal, fine collimation, low kV/high mA, scintillation counter

Of the interferences measured on the ARL 8480, yttrium $K\beta_{1,3}$ (1% of $Y_2O_3 \equiv 0.06\%$ of Nb_2O_5) and zirconium $K\alpha$ (1% of $ZrO_2 \equiv 0.0006\%$ of Nb_2O_5) occurring in the first order are worth noting; that from barium $K\alpha$, in the second order, has no significant effect. If any of these interferences are of significance, a LiF 220 crystal should be used.

7.3.22 Molybdenum Trioxide (normally reported as MoO₃)

Molybdenum is rarely found in traditional ceramic materials or products, except in very small trace amounts. In those specific materials where it is a deliberate addition it is usually present in large amounts.

$L\alpha$ line, Ge crystal, coarse collimator, low kV/high mA, argon/methane flow proportional counter

At its normal low levels the determination of Mo is not usually called for. As a major content, the choice of line falls on the $L\alpha$ with a Ge crystal, coarse collimation, low kV/high mA and a flow proportional counter in the first order. The line unfortunately suffers from a massive interference from sulphur $K\alpha$ (1% of $SO_3 \equiv 0.35\%$ of MoO_3) and smaller ones from zirconium $L\gamma_1$ (1% of $ZrO_2 \equiv 0.005\%$ of MoO_3) and lead $M\alpha$ (1% of $PbO \equiv 0.007\%$ of MoO_3). There is also a weaker interference from some tungsten M lines which could be of relevance in certain circumstances.

For trace levels the $K\alpha$ line is better measured using a LiF 220 crystal, owing to the nearby zirconium $K\beta_{1,3}$ lines (1% of $ZrO_2 \equiv 0.004\%$ of MoO_3 using a LiF 220 crystal). Uranium also causes an interference from its $L\beta_3$ line, but this, again, is reduced by using LiF 220.

It should be noted that molybdenum also causes small to medium interference problems on several common constituents such as silica, magnesia, soda, phosphorus pentoxide, chromium sesquioxide, manganese oxide, zirconia, lead oxide and sulphur trioxide, which may need to be evaluated if circumstances demand it.

7.3.23 Cadmium Oxide (normally reported as CdO)

Cadmium, like so many of the elements discussed in this particular section, is rarely found in ceramic materials as a natural component except at very low trace levels. It is normally found as an additive, usually to yield red, orange or yellow colours, the addition often being made as cadmium sulphoselenide (although not necessarily as a stoichiometric compound). The colour is often diluted, commonly with barium sulphate. During firing, it is suspected that some of the selenium and, presumably, the sulphur is lost to the atmosphere. One of the major problems with cadmium-bearing colours is that they tend to have poor resistance to acids and thus their use is diminishing, owing to the danger of contamination of food. The use of cadmium as a colour for vitreous enamels, for example, has virtually ceased, but recent developments whereby the cadmium compound is 'encapsulated' in a zirconia matrix have gone a long way to ensuring their safe use. This sort of technology needs to be borne in mind since, if cadmium proves to be present in the sample, any of the other above-mentioned elements may also be present and should be sought.

$L\alpha$ line, PE crystal, coarse contamination, low kV/high mA, flow proportional counter

The PE crystal is to be preferred to Ge because of its greater sensitivity. If a tube with a rhodium target is used, it should be noted that there is a nearby rhodium $L\gamma_1$ line. If the cadmium $K\alpha$ line is used there will also be problems due to the rhodium $K\beta_{1,3}$, necessitating the use of a LiF 220 crystal.

7.3.24 Stannic Oxide (normally reported as SnO_2)

Tin is rarely found naturally in ceramic materials, except in impure zircon sands, but is most frequently found as a deliberate addition to act as an opacifier in glasses, glazes and vitreous enamels. It is normally added as stannic oxide at levels usually between 5 and 10% of SnO_2, but has tended to be replaced by a combination of zinc and zirconium oxides as a cheaper alternative. The analysis of tin oxide (SnO_2) itself is occasionally required.

$L\alpha$ line, LiF 200 crystal, fine collimation, low kV/high mA, flow proportional counter

Interferences have been noted from antimony L_n (1% of $Sb_2O_3 \equiv 0.007\%$ of SnO_2) and potassium $K\alpha$ (1% of $K_2O \equiv 0.002\%$ of SnO_2).

7.3.25 Antimony Trioxide (normally reported as Sb_2O_3)

Antimony is normally found only in low trace amounts, except where there is deliberate addition, usually in colours.

$L\alpha$ line, LiF 200 crystal, fine collimation, low kV/high mA, flow proportional counter

Interferences arise from potassium $K\beta_{1,3}$ and $K\beta_5$ (1% of $K_2O \equiv 0.02\%$ of Sb_2O_3) and from the more distant but very intense calcium $K\alpha$ (1% of $CaO \equiv 0.002\%$ of Sb_2O_3).

7.3.26 Iodine (normally reported as I)

Iodine, which occurs normally at very low levels and is also very volatile, is rarely called for, and would not normally be attempted by the fused, cast bead method. Its determination is only attempted in order to monitor residual amounts in the bead if lithium iodide or iodate has been used as a releasing agent.

$L\alpha$ line, LiF 200 crystal, fine collimation, low kV/high mA, flow proportional counter

Iodine $L\beta_{2,15}$ has an interfering effect on titanium $K\alpha$ that may be significant if low level contents of titania are being determined.

7.3.27 Caesia (normally reported as Cs_2O)

Caesium, like rubidium, is normally found only in trace amounts except possibly in materials such as feldspars.

$L\alpha$ line, LiF 200 crystal, fine collimation, low kV/high mA, flow proportional counter

There are interferences from cerium Ll (1% of $CeO_2 \equiv 0.02\%$ of Cs_2O) and barium Ln (1% of $BaO \equiv 0.001\%$ of Cs_2O).

7.3.28 Barium Oxide (normally reported as BaO)

Barium can occur at a range of contents. It is found as a trace to minor impurity in a wide variety of materials; clays, for example, may contain up to about 0.1% of BaO. Similarly, small amounts will be found as impurities in calcium and strontium minerals; here the level may rise to more than 1% of BaO. Glazes, glasses, vitreous enamels, etc., often have barium as a component with levels of content up to several per cent. Finally, there are the materials in which barium may be a major content, either man-made or natural. Barium titanate (the type name covering a wide range of compositions) is an example of the former, and barytes or witherite typify the latter. Barium carbonate, often the precipitated material, is commonly used to suppress soluble sulphates in brick-making.

$L\alpha$ line, LiF 200 crystal, fine collimator, low kV/high mA, argon/methane flow proportional counter

The dominant problem with the determination is the overlap arising from the titanium $K\alpha$ line on the most sensitive barium $L\alpha$ line. The problem is worse than the reverse effect, i.e. barium on titanium, because the barium $L\alpha$ line is less sensitive than the titanium $K\alpha$ line. The optimum settings for the determination are fine collimation, a flow counter and a reasonably wide pulse-height window. The choice of line and crystal is governed by circumstances. The

ELEMENT LINE SELECTION

Table 7.7. The effect of line and crystal on intensity etc. for barium

Line	$L\alpha$	$L\beta_1$	$L\beta_2$	$L\beta_2$	$L\beta_2$
Crystal	200	200	220	200	220
2θ	87.17	79.26	128.78	73.33	115.18
Relative count rate	100	61	40	25	12
Detection limit (10 s)	0.005	0.008	0.01	0.03	0.05
Detection limit with 5% TiO$_2$ (10 s)	0.01	0.01	0.02	—	—
Interference (apparent % BaO) due to presence of:					
1% TiO$_2$	0.037	0.006	0.006	—	—
1% CeO$_2$	—	0.69	0.48	0.02	—
1% La$_2$O$_3$	—	0.002	—	0.25	0.13
1% Nd$_2$O$_3$	—	0.002	—	0.14	0.07
1% V$_2$O$_5$	—	0.007	0.006	—	—

best solution is to use a LiF 220 crystal as for titania. However, at the time of writing, manufacturers do not produce a spectrometer that will reach the required angle of 154.17° (barium $L\alpha_2$). Obviously, some effort on the manufacturers' part in this respect would be welcomed by ceramic and geological analysts alike. A curved LiF 200 crystal used with a fine slit can reduce this interference, with a penalty of lower sensitivity. Another solution would be to use the weaker $L\beta_1$ or $L\beta_2$ lines, but this again is fraught with difficulty. Intensities, line overlap and detection limit data are tabulated below for various line and LiF crystal combinations. The measurements were carried out using an ARL 8480, but the relative effects should hold good for other spectrometers (Table 7.7).

Table 7.7 shows that all combinations suffer some interferences. The $L\beta_1$ line is interfered with by the titanium $K\beta_{1,3}$, cerium $L\alpha_1$, lanthanum $L\alpha_1$, neodymium Ll and Ln and the vanadium $K\alpha$ lines. The barium $L\beta_2$ line is interfered with by the cerium $L\beta_1$, lanthanum $L\beta_3$ and neodymium $L\alpha$ and Ln lines.

Fortunately, in most ceramic materials the TiO$_2$ content does not exceed 5%. It can be seen that the detection limit for the barium $L\alpha$ line in the presence of 5% of TiO$_2$ is the same as that of the $L\beta_1$ line with no titania present, and 3 times better than the $L\beta_2$ line, again with no titania present. Therefore, for most applications the $L\alpha$ line is the one of choice because of its greater sensitivity. In the case of titania and titanium silicates, cerium, lanthanum and neodymium are rarely present, so that the $L\beta_2$ line gives the greatest freedom from titanium interferences. If there is any possibility of the presence of these elements it is as well to carry out a qualitative scan to check for rare earths. Some technical ceramics are deliberately doped with rare earths and titania, and therefore, they must be sought in such materials. In the presence of rare earths, the best choice from $L\beta_1$ with LiF 200, $L\beta_1$ with LiF 220 or $L\beta_2$ with LiF 200 must be made.

It should be noted that the $L\beta_1$ line with LiF 220 is more sensitive than the $L\beta_2$ with LiF 200. The $L\beta_2$ line with LiF 220 gives only one tenth the sensitivity of the $L\alpha$ line on LiF 200, and should be used only where there is no alternative solution.

In addition to the titanium peaks, there is a potential problem on Ba $L\alpha$ from a scandium line, $K\beta_{1,3}$, only 0.14° away (LiF 200). For a rhodium target tube, 1% of $Sc_2O_3 \equiv 0.16\%$ of BaO on an ARL spectrometer and 0.08% on a PW 1606 with curved crystal optics. This would be very serious if a scandium tube is used, unless some precautions are taken. One solution is to use an aluminium filter, or a dual target scandium/molybdenum tube set on molybdenum. Because of the poorer exciting conditions for barium and strontium using these precautions, the lines are less broad, leading to a smaller line overlap of titania on barium (personal communication, S. Davies, Philips, Cambridge). Barium and titania also have large α-corrections on each other, so that careful thought must be given as to how iterative line and α-corrections are applied by the computer, especially where both are present in major amounts, e.g. in the case of barium titanates. This order of correction of line and α-corrections can be critical.

7.3.29 Rare Earths

These are widely distributed, but usually at extremely low trace levels. Their determination is normally required in the ceramic field only when there has been a deliberate addition of a 'pure' rare earth oxide or the sample is a mineral in which rare earths may be expected in significant amounts. A typical case of the latter is zircon beach sand, which may contain appreciable amounts of rare earths; these will then, of course, be found in any preparations containing the mineral. Generally, however, this situation arises at a level to concern the analyst only when the relatively pure zircon raw material usually used is in abnormally short supply. Commonly, in zircon, the oxides of lanthanum, cerium, praseodymium, neodymium and samarium are found at minor levels down to 0.01%. The other rare earths are not normally encountered at levels worth determining (ceramically) when occurring naturally. Deliberate additions are usually made of a single 'pure' oxide. Infrequently, apart from the above, gadolinium and ytterbium may be found as single additions.

When analysing mixtures of the rare earths, lines have to be chosen so as to achieve minimum overlaps rather than to give the best sensitivity. Interferences from caesium and barium may also prove problematical. The above considerations do not normally apply to the determination of an added single oxide, where maximum sensitivity may usually be sought.

Another point to note is that, within the range of wavelengths covered for the rare earths, optimum power settings change from low to high kV. For single oxide types the optimum power setting for that element may be used, but where

ELEMENT LINE SELECTION

all of the above group are to be measured it is probably advisable to retain the lower power setting throughout.

7.3.29.1 Lanthanum Oxide (normally reported as La_2O_3)

$L\alpha$ line, LiF 200 crystal, fine collimation, low kV/high mA, flow proportional counter

The chief problem lies in the interference of caesium $L\beta_4$ (1% of $Cs_2O \equiv 0.09\%$ of La_2O_3), but this is more hypothetical than real, in that caesium contents are normally very low when the determination of lanthanum is called for.

It should also be noted that lanthanum interferes with the chromium $K\alpha$ line.

7.3.29.2 Cerium Oxide (normally reported as CeO_2)

$L\alpha$ line, LiF 200 crystal, fine collimation, low kV/high mA, krypton sealed counter

The biggest problem is interference from barium $L\beta_1$ (1% of $BaO \equiv 0.16\%$ of CeO_2). Normally, BaO is not a problem in traditional ceramics, but difficulty can arise in materials such as barium titanates and bastnasites. In the former the $L\beta_1$ line can be used, but in the latter neodymium $L\alpha$ would interfere. If barium and neodymium are both present, then the $L\alpha$ line using LiF 220 is to be preferred, the disadvantage being a 40% loss of sensitivity. The interference from barium is, however, reduced to about two thirds of the effect suffered using LiF 200.

7.3.29.3 Praseodymium Oxide (normally reported as Pr_6O_{11})

$L\beta$ line, LiF 200 crystal, fine collimation, high kV/low mA, krypton sealed counter

Except when the element is found in the absence of lanthanum, the $L\beta_1$ line needs to be used because of the large interference of lanthanum $L\beta$ on praseodymium $L\alpha$. Small interferences have been noted on the praseodymium $L\beta_1$ due to chromium $K\alpha$ and lanthanum $L\beta_7$. For reasons of sensitivity a krypton sealed counter is preferable to an argon flow counter.

There are third order reflections in this wavelength region, notably that due to uranium $L\beta$ (1% of $U_3O_8 \equiv 0.006\%$ of Pr_6O_{11}). In circumstances where this could be a problem, a flow counter should be used.

If a tube with a chromium target is in use, the chromium $K\beta$ line, more than a degree away, still causes excessive background, necessitating the use of an aluminium filter.

It should also be noted that praseodymium interferes with the chromium $K\alpha$ (1% of $Pr_6O_{11} \equiv 0.005\%$ of Cr_2O_3) and the manganese $K\alpha$ lines (1% of $Pr_6O_{11} \equiv 0.01\%$ of Mn_3O_4), both figures derived from the Telsec TXRF.

7.3.29.4 Neodymium Oxide (normally reported as Nd_2O_3)

$L\beta_1$ line, LiF 200 crystal, fine collimation, high kV/low mA, krypton sealed counter

The $L\alpha$ line is to be avoided unless the sample is free from other rare earths because of interferences from cerium $L\beta_1$ and, to a lesser extent, lanthanum $L\beta$ lines. A krypton sealed counter is normally preferable to an argon flow counter in terms of sensitivity. If uranium is likely to be present, there is a third order uranium $L\beta_1$ line that drops into the krypton detector (1% of $U_3O_8 \equiv 0.001\%$ of Nd_2O_3). This effect is eliminated by the use of a flow counter. Fortunately the fourth order rhodium $K\beta_1$ line does not cause a problem with the krypton detector.

As with the earlier mentioned praseodymium $L\beta$ line, chromium causes a background problem if the tube has a chromium target via the chromium $K\beta_{1,3}$ line, again necessitating the use of an aluminium filter.

It should be noted that neodymium interferes with the Mn Kα line (1% of $Nd_2O_3 \equiv 0.007\%$ of Mn_3O_4, on the Telsec TXRF).

7.3.29.5 Samarium Oxide (normally reported as Sm_2O_3)

$L\beta_1$ line, LiF 200 crystal, fine collimation, high kV/low mA, krypton sealed counter

This oxide is usually found at very low content levels, so that line overlap interferences from other rare earths on the $L\alpha$ line can be very serious (Table 7.8). It is therefore advisable to use the $L\beta$ line.

For the samarium $L\beta_1$ line, a krypton sealed counter is more sensitive than an argon flow counter. The third order peak due to niobium $K\beta_{1,3}$, although high (1% of $Nb_2O_3 \equiv 0.02\%$ of Sm_2O_3), is normally of little concern, because of the improbability of finding both oxides together in the same sample.

7.3.29.6 Gadolinium Oxide (normally reported as Gd_2O_3)

$L\alpha$ line, LiF 200 crystal, fine collimation, high kV/low mA, krypton sealed counter

Table 7.8. Interferences on samarium

Element	Line	% Error due to 1% of interfering oxide on samarium $L\alpha$
Cerium	$L\beta_{2,15}$	0.04
Praseodymium	$L\beta_3$	0.01
Neodymium	$L\beta_1$	0.008

ELEMENT LINE SELECTION

This oxide is usually found essentially on its own. Even in the presence of other rare earths, the level of samaria is usually so low that little difficulty arises from samarium $L\beta_1$. A krypton sealed detector is usually to be preferred, especially as no third order interferences are experienced.

It should be noted that gadolinium is responsible for small interferences on phosphorus $K\alpha$ and chromium $K\alpha$.

7.3.29.7 Ytterbium Oxide (normally reported as Yb_2O_3)

$L\alpha$ line, LiF 200 crystal, fine collimation, high kV/low mA, krypton sealed counter

Ytterbium is usually found as a deliberate addition in ceramic materials, effectively free from other rare earths, so that the $L\alpha$ line can be used. There are some first order interferences from non-rare earth oxides that could potentially cause problems, but in practice seldom do. These are due to tungsten Ll (1% of $WO_3 \equiv 0.03\%$ of Yb_2O_3) and nickel $K\alpha$ (1% of $NiO \equiv 0.06\%$ of Yb_2O_3).

Second order reflections are just as significant when a krypton sealed counter is in use as they would be with a flow proportional counter in this region, so that the former is to be preferred because of its greater sensitivity. Rubidium $K\beta_{1,3}$ in the second order does not cause problems, but yttrium $K\alpha$ and lead $L\gamma_1$ do. The degree of interference on each of the two counters is shown in Table 7.9.

No significant third order reflections have been found.

7.3.30 Hafnia (normally reported as HfO_2)

Hafnia is hardly ever found, except in the presence of 40–50 times as much zirconia, so that its determination is only considered under such circumstances.

$L\beta_2$ line, LiF 200 crystal, fine collimator, low kV/high mA, flow counter

The complex and involved reasoning behind the choice of best parameters for this line would almost require a chapter in itself. However, the essence of it is set down here both as a record and to avoid the necessity of the reader having to go over the same ground in other cases.

Table 7.9. Interferences on ytterbium on two counters (as % Yb_2O_3)

Interferent	Counter	
	Argon flow	Krypton sealed
1% Y_2O_3	0.0017	0.0025% Yb_2O_3
1% PbO	0.011	0.007 % Yb_2O_3

The apparently obvious line to choose is the $L\alpha$ line, but this is interfered with by the nearby zirconium second order $K\alpha$ line, and as HfO_2 is found in the presence of about fifty times as much ZrO_2 the latter line will thus be a strong one. One might expect the pulse-height discrimination to remove this, but in the case of argon and krypton the escape peaks cause the interference. In addition, the resolution of the scintillation counter is about three times poorer than that of the krypton counter, creating even more interference.

Measured on an ARL 8480 using a LiF 200 crystal with fine collimation, the apparent HfO_2 concentration shown by 65% of ZrO_2 was 0.17% for an argon flow, 14.3% for a Kr sealed and 10.1% for a scintillation counter. The sensitivities were similar for all three counters, a proportion of 6:7:6 Ar:Kr:Scint. Even with the flow counter the background effect is unacceptable, bearing in mind that the levels of hafnia are only about 1.4%.

Using a LiF 220 crystal the sensitivities drop to about a third but the line overlaps are equivalent to 0.11, 7.3 and 0.93% of HfO_2 for the three counters respectively.

A better remedy is to drop the kV setting to less than the 60 kV normally used. This reduces the sensitivity of the zirconium $K\alpha$ line more than the hafnium $L\alpha$ line. Changing from the settings of 60 kV/50 mA to 30 kV/100 mA, the sensitivities drop to about two thirds but the line overlap for the three counters reduces (for 65% of ZrO_2) to 0.014, 6.8 and 2.7% for argon, krypton and scintillation counters.

The Ge crystal will eliminate second order reflections, so it would be an obvious alternative method of overcoming this second order zirconium $K\alpha$ overlap. Unfortunately the zirconium in the sample very efficiently excites the germanium K lines. It is not possible to remove these fully by pulse-height analysis, and to make matters worse the escape peaks fall even closer to the pulse height window set. The ARL 8480 spectrometer has not been calibrated at this low level using the Ge crystal; however measurements taken on the Philips PW 1410 showed that 65% of ZrO_2 gave an interference equivalent to 0.41% of HfO_2 with a flow proportional counter and 0.0645% with a scintillation counter.

To summarize the arguments so far, for the hafnium $L\alpha$ line a LiF 200 crystal with a flow counter provides the best combination of crystal and detector to reduce zirconia interference, but a low voltage setting, e.g. 30 kV/100 mA, has also to be used to bring this level of interference to an acceptable level (0.014% of $HfO_2 \equiv 65\%$ of ZrO_2). The reduced kV reduces sensitivity to about two thirds of that of the higher settings. The LiF 220 crystal is not a good option.

The hafnium $L\beta_1$ line is reinforced by a nearby $L\beta_3$ line and a still weaker $L\beta_6$ line. There is again a second order interference from zirconia here, but it is caused by the much weaker $K\beta_{1,3}$ line. This line is better separated from the hafnium $L\beta$ line than is the zirconium $K\alpha$ line from the equivalent hafnium $L\alpha$ line. At a setting of 60 kV/50 mA the intensity of this hafnium line ($L\beta_1$)

ELEMENT LINE SELECTION

is about 90% of that of the $L\alpha$ line at a similar power setting, and hence about 1.4 times more sensitive than the $L\alpha$ line at 30 kV/100 mA. The flow counter gave the lowest interference for the Hf $L\alpha$ line, and the interference from 65% of ZrO_2 measured on the $L\beta$ line using a PW 1410 was only 0.0007%, much better than the best settings on the $L\alpha$ line.

The PW 1410 and similar instruments have the scintillation counter mounted behind the flow counter; this seems to lead to reduced sensitivities on the scintillation counter but does permit the flow counter to be used at low angles. The ARL 8480 has two or three counters mounted alongside each other. As the flow counter is usually needed at high angles (e.g. for aluminium $K\alpha$ on PE or titanium $K\alpha$ on LiF 220) it is placed at the high angle side of the counter mounting. This means that it cannot be used at an angle of 38.89° for hafnium $L\beta$, and thereby forces an alternative course of action. Using the scintillation counter on the ARL 8480, 65% of ZrO_2 gave an interference of 0.009% of HfO_2; the krypton counter gave one of 0.43%. A xenon sealed counter on the Telsec TXRF gave an overlap of 65% of $ZrO_2 \equiv 0.0095\%$ of HfO_2. The LiF 220 crystal is again no advantage, but if really necessary the overlap could be effectively eliminated with a low mA setting.

To summarize, the best way of eliminating zirconium interference on hafnium is to measure the $L\beta_1$ line with a LiF 200 crystal, a flow counter and fine collimation, preferably at lowish kV and high mA. For a PW 1606 with a LiF 200 crystal and an argon/methane flow counter at 40 kV/75 mA, the line overlap of ZrO_2 on hafnium $L\beta_1$ was 65% of $ZrO_2 = 0.0097\%$ of HfO_2. Dropping the kV to 30 effectively eliminated the interference completely. If an ARL 8480 is used then the same conditions with a scintillation counter or preferably a xenon sealed counter give adequate reduction of the overlap effect. The pulse-height window should be set as tight as possible to cut out second order energies and the escape peaks.

It has not been found necessary, in practice, to measure a background line in normal analysis. In fact, the spectrum around the hafnium $L\beta_1$ is a positive minefield of interfering lines that would make measuring a background extremely difficult. If one is needed, the only safe place to measure is some distance away at 35° 2θ (LiF 200). This background position could be used as a general one for many elements in this area of the spectrum.

7.3.31 Tantalum Oxide (normally reported as Ta_2O_5)

$L\alpha$ line, LiF 220 crystal, fine collimation, high kV/low mA, krypton sealed counter

There are some risks of interference here, but for many purposes a LiF 200 crystal could be used. If niobium, copper or uranium is present in the sample its effect can be approximately halved, albeit with a three-fold decrease in

Table 7.10. Line overlaps on tantalum (as % Ta_2O_5)

Oxide	Line	Order	Effect of 1% of oxide
Nb_2O_5	$K\alpha$	2	0.01
CuO	$K\alpha$	1	0.005
U_3O_8	$L\beta_2$	2	0.003

sensitivity, by using a LiF 200 crystal. Overlaps have been found, as shown in Table 7.10.

7.3.32 Tungstic Oxide (normally reported as WO_3)

Apart from its very useful characteristics of hardness and wear resistance, the fact that the element rarely occurs naturally in ceramic samples makes tungsten carbide a very convenient material to use for mortars for grinding the sample in the laboratory. Thus, this element will be introduced in small amounts as contamination during grinding. The level of content is normally low, and with many of the softer types of material its presence would not cause noticeable error. Nevertheless, with harder types of material, and there are many, contamination can reach identifiable proportions and occasionally enough for it to be essential to correct for its presence. For accurate analysis it is therefore at least desirable and often necessary to monitor WO_3 in the fused, cast beads in order to establish the degree of contamination. The WO_3 content so determined is then used to correct the total analysis for the dilution factor its presence introduces: the weight of sample in the bead is less than that intended due to the weight of tungsten carbide introduced. It is also necessary to correct the measured loss on ignition for the gain in weight due to the reaction:

$$2WC + 5O_2 \rightarrow 2WO_3 + 2CO_2$$

The mathematical equations and computer algorithms for these corrections are given later when the method is being described (Chapter 9).

$L\alpha$ line, LiF 220 crystal, fine collimator, high kV/low mA, scintillation, flow plus scintillation or xenon sealed counter

Because of the foregoing, it can be seen that it is imperative, in the determination of WO_3, to ensure that the signal being measured by the XRF is due solely to the tungsten and not to some line overlap. Sensitivity is of lesser importance because errors due to poorer counting statistics will only result in secondary errors on the determined tungsten carbide figure. These errors will virtually disappear when the calculations for correction for the presence of tungsten are applied to the sample. The errors are certainly going to be less disastrous than applying corrections for an interference caused by the presence

ELEMENT LINE SELECTION

of line overlap from an element whose content could possibly vary appreciably. Therefore, tungsten is best determined using the $L\alpha$ line in combination with a LiF 220 crystal and fine collimation. The detector should be a scintillation counter, a flow plus scintillation counter or a xenon sealed tube. The X-ray spectrum around the tungsten $L\alpha$ line is fairly abundant with interfering lines and, because scintillation counters are used, higher order lines can also be a problem. For this reason the xenon sealed counter with its better resolution would be preferable if it is available. The abundance of lines in this area also make the choice of an off-peak background very difficult and, for this reason, it is better not to measure a background but to apply mathematical corrections, treating changes in background as positive or negative line overlap corrections. Even applying the precautions given above, there are some notable line overlap effects that cannot be excluded and have to be corrected for. These are given below with an indication of their magnitude as measured on a Telsec TXRF with a xenon counter.

Zinc $K\alpha$ 1% of ZnO ≡ 0.015% of WO_3
Nickel $K\beta_{1,3}$ 1% of NiO ≡ 0.004% of WO_3

If rare earths are expected the ytterbium $L\beta_1$ line can also cause problems.

1% of Yb_2O_3 ≡ 0.5% of WO_3

Although rare earths are not normally sought, their presence can be indicated on the chromium $K\alpha$ and manganese $K\alpha$ lines, as has been discussed. Occasionally ytterbium is deliberately added to ceramics. Yttrium also has a small line overlap from a second order $K\beta_{1,3}$ peak (1% of Y_2O_3 ≡ 0.008% of WO_3, as measured on an ARL 8480).

7.3.33 Lead Monoxide (normally reported as PbO)

As with a number of other oxides, the determination of lead oxide is required at several levels. Starting at the lowest level, trace amounts of lead need to be measured both for ceramic investment casting materials and for environmental purposes. Next comes its appearance as a minor content in glazes, glasses etc; contents of lead in such materials range, in general, from 2% up to, say, 25% of PbO. Major contents can be required in some lead frits, e.g. lead bisilicate (63% of PbO), and even higher in lead-bearing ores such as galena. Thus, determinations of lead can be called for at any level of content from the detection limit of a few parts per million through to over 80% of PbO.

$M\alpha$ line, Ge crystal, coarse collimator, low kV/high mA, argon/methane flow proportional counter

7.3.33.1 Lead $L\beta_1$

It is clear that a single calibration curve cannot be expected to cover the whole of this range, any more than one set of XRF instrument settings. For reasons given previously, e.g. for ZrO_2, the longer wavelength line, in this case the lead $M\alpha$, is preferable for the analysis of glazes etc. At low levels the more sensitive $L\beta$ line is used. The $L\alpha$ line is ruled out because of the arsenic $K\alpha$ line, which is right on top of it. It is desirable to compensate for changes in background from one sample to the next by measuring an off-peak position at $+0.7°$ using a LiF 200 crystal.

7.3.33.2 Lead $M\alpha$

Apart from the obvious reason of temperature stability, the Ge crystal is preferable to PE because it provides a better resolution from the nearby sulphur $K\alpha$ and molybdenum $L\alpha$ lines. For a small loss of about 20% in sensitivity the improved resolution is quite dramatic.

One point to bear in mind is that the lead $M\alpha$ line is interfered with by third order iron $K\beta$. This interference is worse on the body centred cubic Ge crystal than on PE; although the Ge crystal eliminates even ordered reflections, the odd orders are enhanced (cf. Co on S). It is necessary to apply a narrow pulse-height window in order to minimize the interference from this ubiquitous element. For an ARL 8480, a change from a window of 4 to one of 2 reduces the interference due to iron by half, with very little loss on sensitivity. Below are tabulated (Table 7.11) some important interferences on lead $M\alpha$ measured on an ARL 8480 with coarse collimation and a pulse height window of 2 over a threshold of 5.

It can be seen from Table 7.11 that the Ge crystal is greatly superior in resolution to PE for the first three interferences. As explained above, in the case of iron, the interference is worse but tolerable unless very low lead contents are being sought, in which case the $L\beta$ line would be used for other reasons. This effect of iron was noticeable on an ARL 8480 but too small to be measured

Table 7.11. Interferences on lead $M\alpha$ line (as % PbO)

Interferent			Crystal	
	Order	Line	PE	Ge
1% SO_3	1	$K\alpha$	0.382	0.094
1% MoO_3	1	$L\alpha$	0.140	0.018
1% Bi_2O_3	1	$M\alpha$	0.015	0.005
1% Fe_2O_3	3	$K\beta_{1,3}$	0.001	0.002
1% ZrO_2	1	$L\gamma_1$	0.002	0.004

ELEMENT LINE SELECTION 143

on a PW 1606. The increased interference of zirconium with the Ge crystal is odd, as one would expect the zirconium $L\gamma$ line to be better resolved. It can only be assumed that part of the interference is due to an increased background effect using Ge that is not so marked with PE. If absolutely necessary, these interferences can be greatly reduced by using fine collimation. However, the fourfold drop in sensitivity of this line is normally too high a price to pay. In practice coarse collimation has been found to suit all those analytical situations encountered where the $M\alpha$ line is used in preference to the $L\beta$ line.

7.3.34 Bismuth Oxide (normally reported as Bi_2O_3)

From experience it appears that Bi_2O_3 is either added in substantial amounts or is at so low a level that its determination by XRF is scarcely practical. The latter is particularly true in the analysis of ceramics used for investment casting, where specification levels of about 1 ppm maximum mean that either AAS or ICP is required; even then, hydride generation techniques or a graphite furnace are usually needed.

$M\alpha$ line, Ge crystal, fine collimation, high kV/low mA, scintillation counter

In general the $M\alpha$ line has been preferred to the $L\alpha$ line for similar reasons to those discussed under the determination of lead.

7.3.35 Thoria (normally reported as ThO_2)

$L\alpha$ line, LiF 200 crystal, fine collimation, high kV/low mA, scintillation counter

This oxide is, like so many of these unusual elements, only found at low levels, so that the $L\alpha$ line is preferred. The most significant nearby interference is due to bismuth $L\beta_1$ (1% of $Bi_2O_3 \equiv 0.7\%$ of ThO_2). This combination of bismuth with thorium is unlikely to occur in practice, but if it does the $L\beta$ line should be used. The next most significant interference is from lead $L\beta_1$ (1% of $PbO \equiv 0.01\%$ of ThO_2). If the lead interference problem ever arises, which appears unlikely, the LiF 220 crystal should be used.

7.3.36 Uranium Oxide (normally reported as U_3O_8)

$L\alpha$ line, LiF 200 crystal, fine collimation, high kV/low mA, scintillation counter

This line is used for low contents. A background measurement at $-0.5°\ 2\theta$ is advisable with a LiF 200 crystal.

$M\alpha$ line, Ge crystal, coarse collimation, low kV/high mA, flow proportional counter

This is the line normally used for higher contents such as those in ores.

$M\beta$ line, LiF 200 crystal, coarse collimation, low kV/high mA, flow proportional counter

This line is more intense than the $M\alpha$ line, as it can be measured on the more sensitive LiF 200 crystal. The sensitivity is about three times greater than that of the $M\alpha$ line measured on a Ge crystal, and compares very favourably with that for the $L\alpha$ line. Unfortunately the $M\beta$ line suffers very severely from interferences from potash. To achieve better sensitivity than can be obtained from the $M\alpha$ line, coarse collimation must be used. This gives rise to overlap from potash $K\alpha$, such that 1% of K_2O is equivalent to 0.4% of U_3O_8, measured under these conditions. The use of fine collimation reduces the magnitude of the line overlap; this will, of course, differ from instrument to instrument. On an ARL 8480 the overlap was reduced to an equivalent of 0.12% with a sixfold loss of sensitivity compared with that of the $M\alpha$ line. This illustrates the merits of juggling with crystals and collimation. For practical purposes the $M\beta$ line is restricted to samples in which the potash content is close to zero.

8 The Standard Procedure

8.1 INTRODUCTION

Over the whole range of types of sample, there are quite a number of deviations from the generic method; these are for the most part merely modifications related to the nature of the material being analysed. The details of such changes, where needed, are described in the appropriate section of the book (Chapters 12-19). However, to provide a complete coverage of the operations, it is necessary to describe in detail the standard procedure for the analysis of uncomplicated samples, particularly as these are likely to form the main usage. In this description it is assumed that the analyst will carry out a loss on ignition determination as part of the analysis, rather than individual determinations of water, carbon dioxide, etc. This is not done because it is assumed that the interest will lie in the analysis of ceramic materials, but rather that to use the method to the best advantage an ignited sample is required. As previously discussed, the sample has to be ignited so that dilution in the flux can be exact. It is true that corrections can be applied, enabling figures to be obtained without recourse to other than a raw sample and a knowledge of the loss on ignition figure, but such analyses are more prone to error than those in which an ignited sample is used. Furthermore, if a raw sample is used, the deficit in analytical total (of the determined oxides) has been known to be used to calculate a 'loss on ignition' (more correctly loss on fusion). If this is done no true analytical total is achieved, so that one of the built-in accuracy checks is lost.

The description of the method is given in much the same format as used by BSI, as this is used by Ceram Research, through their Working Groups, in their role as providers of the basic procedures submitted for adoption as standard methods. This has become normal practice in the writing up of methods, and has been the case for more than thirty years. The method used as an example is that now adopted as the British Standard Method for the Analysis of Aluminosilicate Materials (BS 1902: Part 9, Section 1). To the formal description of the procedure has been added, hopefully, enough know-how to enable a reasonably experienced analyst, or a young analyst under supervision, to carry out reliable analyses of commonly met materials. It is realized that this involves some duplication but as the reference may be at different times and for different purposes it is thought to be justified. The chapter is primarily concerned with the procedure to the point of presentation of the bead to the instrument; the details of calibration and calculation of results are dealt with in later chapters (Chapters 9 and 10 respectively).

The British Standard Method (BS 1902: Part 9, Section 1) permits a number of alternative procedures. These have been included in the BS method as they represent different current practices, all of which have been shown to be acceptable. In this treatment, only those alternatives that have found most favour with the authors are included in detail.

The most important thing is that, having selected the details of the method most suitable to a given laboratory, these must be followed rigorously, both in calibration and analysis. If it is desired to change ANY detail it is essential to start again, calibrating and analysing by the modified method. The only exception to this is if it can be proved conclusively over the whole range of materials to be analysed that the change has no effect on the results.

The information given in normal type is that pertinent to the formal description of the method. The remaining information, given in smaller type, is additional information that is best described as 'know-how' and will hopefully be useful to analysts not fully conversant with such background. Much of this is a distillation of data discussed elsewhere in the book.

8.2 DETERMINATION OF THE LOSS ON IGNITION

Ignite a 50 mm diameter clean platinum/gold dish to red heat over a gas burner and transfer to a desiccator containing silica gel, followed by the dried (110°C) sample in its corked sample tube. Both sample and dish are allowed to cool to room temperature in the balance room for 20–30 min.

> The exact time depends on the number of samples being cooled in the desiccator; ideally it should be the minimum necessary to cool the contents to room temperature. It is good practice to standardize the time as far as possible so as to avoid possible deviations in temperature or moisture absorption etc.

Weigh the dish, and into it weigh 2 g of the dried, cooled sample. Two alternative approaches exist, either the exact amount of 2.000 g may be weighed out or a dead weight of approximately 2 g.

> Each course has its proponents, as each has advantages. It is legitimate either to weigh the dish plus lid before and after the procedure or simply to weigh the dish alone. The former reduces the risk of moisture etc. pick up during the weighing of the ignited sample, but is clearly less convenient. In any event, the modern balance can be operated so quickly that, in most cases errors from the above source are negligible, making the weighing of the dish without the lid preferable.

THE STANDARD PROCEDURE 147

Partially cover the dish with a lid, still allowing access of air to the contents. Place the sample either in a cold furnace or over a gas burner and progressively raise the temperature to 1000°C or the full heat attainable by the burner, reaching this state in abut 10-20 min.

> The most recent generation of lightweight and microwave furnaces, capable of reaching temperatures of about 1000°C in a few minutes, may have made the former alternative far more practical than was previously the case. However, unless the sample contains high levels of carbonaceous material and/or iron compounds, there is little to be gained from the use of a furnace throughout. The use of gas burners enables a higher throughput unless the laboratory is prepared to purchase a number of rapid heating furnaces. The only advantage of such furnaces lies in the fact that oxidizing atmospheres are more easily maintained, but even so the rate of heating may need to be slowed down to cope with adverse concentrations of carbon and iron. When the sample does contain significant amounts of organic matter, e.g. ball clay or fireclay, great care needs to be taken to avoid raising the temperature above a very dull red heat until all organic material has been burnt out. This point can usually be identified by the removal of the grey colour of the material, and, if the sample contains appreciable amounts of iron, the development of a pink-brown to red tint in the sample. Failure to observe this precaution can result in any of several undesirable occurrences. The carbon may prove extremely difficult to eradicate once it has reached too high a temperature; it is suspected that this may be due to the formation of some silicon carbide. Although this has not been proved, a similar phenomenon has been observed during combustion of filter papers containing precipitated silica during 'wet' analysis. In addition, damage may be caused to the dish by the reduction of iron in the sample to the metallic state, followed by alloying with the metal of the dish. This not only wastes a lot of analyst time in cleaning up the dish but also removes iron from the sample, potentially resulting in an incorrect analysis. Finally, there is a danger that the carbonaceous material itself will attack the metal of the dish, forming carbide and resulting in permanent damage.

After reaching the maximum temperature obtainable with the gas burner, the dish and contents are transferred to the furnaces (1025°C) for 30 min, with the lid still almost covering the dish. If the process has been carried out throughout in a furnace, the sample is similarly heated after attaining the desired temperature.

> An electric furnace is preferable for the final heating, as there is then no danger of any atmosphere less than fully oxidizing. A reducing atmosphere could conceivably arise using gas. This is important, since the higher the temperature the more easily is iron reduced to the metal. In addition the temperature to which the sample is being heated is more easily controlled.
> The determination of loss on ignition on samples falling into this general category may be carried out using a tunnel kiln. Indeed, this applies to many other types of sample where the schedule is suitable. This kiln has been described earlier (Chapter 2). It has the considerable advantages of a very high potential throughput, a controlled and steady heating-up rate and the maintenance of oxidizing conditions throughout. On no occasion while igniting normal-type samples has there been

any indication of a failure to achieve complete oxidation, or any sign of reduced iron. This does not, of course, apply to samples such as chrome-bearing materials, where much of the iron is held within the spinels.

Some types of sample tend to sinter at or below 1025°C. If the degree of sintering is expected to be too severe to allow the sample to be used for bead preparation, it is best to carry out a separate loss on ignition determination. In this case the sample can be weighed into a platinum or, preferably, a porcelain crucible for the loss on ignition determination so as not to 'tie up' a fusion dish. If this is done, a portion of the dried (110°C) sample, with the weight corrected for the loss on ignition, must be weighed out for the fusion. Materials with a low melting point, such as glazes or glasses, or even just samples that may be expected to sinter and adhere to platinum, are best ignited in a porcelain crucible. This crucible, being inexpensive, can then be discarded after use, avoiding the use of analyst time for cleaning up the platinum vessel.

After the ignition, use the lid to provide a complete cover for the dish and transfer the dish, lid and sample to a desiccator containing silica gel; if the sample is likely to absorb carbon dioxide, it may help to add an appropriate absorbent to the desiccant (e.g. Carbosorb). After the cooling time (the same as used earlier when weighing the empty dish), re-weigh the dish and contents.

It is usually unnecessary to repeat the ignition, but, with unfamiliar types of sample or types expected to need extra ignition, a further 30 min should be given. The second figure should check with the first within acceptable limits; 1 mg is reasonable for most purposes, but this figure is a decision for the individual and must depend on requirements and circumstances.

8.2.1 Calculation

If W_1 = weight of empty dish,
  W_2 = weight of dish + sample before ignition,
  W_3 = weight of dish + sample after ignition,
then
  $W_2 - W_1 = W_B$ (weight of sample before ignition)
  $W_3 - W_1 = W_A$ (weight of sample after ignition)
and $\dfrac{W_B - W_A}{W_B} \times 100 =$ loss on ignition (%)

8.3 PREPARATION OF THE BEAD

8.3.1 General Criteria

There are a number of fundamental principles that apply to the method. The first of these does not apply to the whole range of samples discussed in the book, but the remainder do. These include the following.

THE STANDARD PROCEDURE

(a) The ratio by mass of the flux to the sample is 5:1.

(b) The melts produced need to be homogeneous.

(c) The conditions of fusion need to be such that there will be no measurable loss of any determinand from the sample during the process e.g. loss of iron by reduction or alkalis by evaporation.

(d) The conditions of fusion need to be such that any loss of flux is reproducible.

(e) The fusion needs to be conducted in such a manner that the sample is not contaminated in any significant manner.

(f) The beads prepared to be used for presentation to the instrument must be free from blemishes on the chosen analytical surface.

(g) If the top surface of the bead is to be used for analysis, it must be, either convex or flat (i.e. not concave) and symmetrical across any diameter.

It should be noted that in the 'General Criteria' given in the Standard Method above, items (b), (c), (d) and (e) are all counsels of perfection. The criteria are ideals to which the analyst must pay attention, using whatever reasonable precautions can be taken in the context of the work-load. These criteria are to ensure that any errors arising from other relevant sources are minimal and do not noticeably affect the results of the analysis. It is clearly impossible to guarantee complete adherence to these conditions. This becomes clear if one examines the conditions pedantically.

In (b), there can be no certainty that homogeneity has, in fact, been achieved, except by examining the whole bead using, say, scanning electron microscopy (SEM). Moroever, it is a well known fact in the manufacture of glass that the composition of the surface layers may be expected to differ from the composition of the bulk even if only minutely. Slight losses of volatile components can occur from the surface while the glass is still hot but becoming viscous, after pouring. The prime need, therefore, is to make the process reproducible. Adequate swirling should produce a melt homogeneous 'for all practical purposes' and by controlling the casting process carefully, consistent variations in composition from the surface to the bulk can be achieved, so that accuracies of results are unaffected by this slight lack of inhomogeneity with depth.

With respect to (c) the key word is, of course, 'measurable'. Alkalis may be lost to some very slight degree, but if the calibration and analyses are carried out in a sufficiently similar manner, then losses from standard and sample beads should be virtually identical and the results will be correct. The same principle exactly applies to (d).

Finally, dealing with (e), contamination will occur. In fact, it is inevitable. There will be contamination arising from the flux, but this should be slight and acceptable if a suitable flux is used. There will also be contamination from the metal of the dish. Examination of the bead will inevitably reveal traces of both platinum and gold. These are probably so small as to be ignorable, but, in any event, once again the comparison of sample and standard beads, taken reproducibly through the process, should cancel out even this very low level of error.

8.3.2 Fusion

Provided that the portion of the sample used for the loss on ignition has not sintered together, i.e. the sample can be crushed by gentle pressure with a spatula, it is advantageous to use this for preparing the bead.

The weights given below are for the preparation of the more commonly used bead size, viz 35 mm diameter. If a different bead size is to be used, weights of sample and flux will need to be changed.

Weigh 1.500 g of the sample for the fusion process. If the portion of the sample used for the loss on ignition determination is to be used, sample is removed from the dish until 1.500 g remains. Normally this would be carried out by removing successive increments from the dish, transferring them to a small piece of paper (e.g. small ashless filter paper) and transferring back from the paper, if necessary. When this is not possible and a fresh portion of sample has to be used, the weight needs to be calculated so that the weight of ignited sample would be 1.500 g. The required weight of raw sample to yield 1.500 g of ignited sample may be calculated as follows.

If w_r = weight of raw sample, and L = loss on ignition (%), then

$$w_r = \frac{1.500}{[1 - (L \div 100)]}$$

Then, add to the dish containing the sample the equivalent of 7.500 g of dry 4:1 flux. This can either be in the form of freshly dehydrated flux (dehydrated overnight) or by weighing an amount of flux, calculated from a previously determined loss of weight during dehydration, straight from the container to yield 7.500 g of dehydrated flux. Again, if W_r = weight of raw flux, and m = loss of weight of flux on dehydration (%), then

$$W_r = \frac{7.500}{[1 - (m \div 100)]}$$

Mix together the sample and flux carefully and thoroughly in the platinum/gold dish, using a spatula or platinum wire, partially cover the dish with a lid and transfer the dish and contents to the low flame of a gas burner. During a period of 5–10 min, raise the temperature progressively to the full heat of the burner, but with the lid only partially in place, so as to allow ready access of air.

> If unignited sample is being used, this preliminary progressive heating will need to proceed more cautiously to allow time for the burning out of carbonaceous material in the same fashion as would be necessary for carrying out a similar loss on ignition determination. If the flux starts to melt prior to the removal of all potentially reducing material, air can become completely excluded at the reaction point and all the dangers inherent in reducing conditions can arise.

Once the flux has melted, swirl the melt occasionally to ensure both that no unattacked sample is adhering to the sides of the dish, thus escaping the attacking melt, and that adequate homogenizing of the melt occurs.

THE STANDARD PROCEDURE 151

This process is easily carried out using the multiple swirling burner device (Chapter 2). Once the device has been set in motion, attention is needed only in the first few minutes while the temperature is being raised. Once melting of the flux has occurred, the fusion process can be left unattended until it is time to check on the completeness of decomposition.

Transfer the dish and contents to the 1200°C furnace, together with the casting dish and its heat reservoir: a roughly 50 mm square piece of sillimanite batt of about 10 mm thickness will suffice.

8.3.3 Casting of Beads

Beads may be cast by any one of several methods, each of which is described in the Standard Method. These include both the use of a combined fusion and casting dish, and casting inside and outside the furnace. Two procedures are given for the latter, the first involving heating the casting mould over a gas burner and the second using a heat reservoir.

It is, of course possible to cast the bead inside the furnace, as suggested in one of the alternatives in the Standard Method, but manipulation is more difficult and possibly dangerous. From the analyst's point of view it is certainly more uncomfortable. It is equally valid to heat the casting mould on the burner, but maintaining it there in the horizontal plane while casting is more difficult. The method described here appears to be the simplest and safest. With regard to safety, it is essential that operators should wear safety glasses to prevent any possibility of glass being ejected into the eyes by a bead shattering due to thermal shock; i.e. internal stresses caused by rapid changes in temperature after solidification. Also, it is imperative that shoes which completely cover the fore part of the foot (i.e. not open-toed) be worn so that if any molten flux is spilled the foot will not be burnt.

After 5 min, remove the heat reservoir from the furnace and place it on a flat, horizontal, heat-resistant surface, and then transfer the casting mould so that it rests on the heat reservoir. Remove the lid from the fusion dish and remove the dish from the furnace, quickly swirl the contents and pour the melt into the casting mould without delay. A 'rippled' top surface produced in the casting process can lead to erroneous results. In order to avoid this 'rippled' effect, the last portion of the melt should be poured into the mould at a point nearer to the edge of the mould than the centre.

When using top surfaces, in order to maintain a uniform curvature on the top surface, it is necessary to transfer as much of the melt into the casting mould as possible so as to achieve consistent bead weights.

8.3.4 Cooling of Beads

If an air jet is used, transfer the mould to it either while the melt is still quite fluid or after cooling on the flat, horizontal surface until the melt has cooled from red heat, i.e. until it is solid. If the former course is adopted the support over the air jet* must be horizontal when using top surfaces. If no air jet is used, allow the mould to cool on a horizontal surface.

Retain the dish in a horizontal position above the air jet so that the air is directed on to the centre of the base of the mould. When the bead has solidified and released, turn off the air jet. (The progress of release is often indicated by the formation of Newton's rings.)

Note: It may be necessary to encourage the release of the bead by *gently* tapping the casting mould on to a solid surface.

The important points to note in regard to casting and cooling, particularly when the top surface is to be used for measurement, are:

(1) the melt should be poured whilst it is still very fluid, so that it has time to come to rest before becoming viscous. Pouring into the casting mould somewhat off-centre helps.

(2) Rippling appears to arise both from insufficient attention to (1) and from allowing draughts to blow across the surface during cooling. The effect may be due to a combination of local increase in viscosity, or even solidifcation before the rest of the melt, or to stresses caused by temperature differences at various points in the melt. The use of the air-jet helps to even out the latter, but the movement from the original cooling surface to the air-jet should not be done as the melt is setting. Sometimes a similar effect is caused by bad mixing; this can only be cured by reverting to correct practices for carrying out the fusion.

(3) The surface to be measured should be free from flaws, as these will affect the results. Bottom surfaces are more prone to bubbles than top, but experience has shown that both surfaces give equally good results once the necessary skills have been acquired. An individual laboratory will select one technique or the other, after which results from that laboratory will demonstrate that its chosen technique gives better results than the alternative, and very significantly so.

The use of the top surface seems to be simpler and involves less maintenance of the casting moulds. Problems of contamination and partial devitrification have also been experienced when using bottom surfaces.

8.3.5 Automatic Bead Preparation

Automatic bead equipment may be used, but it must satisfy all the requirements of the General Criteria.

---

*A suitable air jet may be derived by using the jet at the base of a Bunsen burner. The burner is connected via the gas pipe to an air flow derived either from a small compressor or, more simply, by allowing water to flow into an aspirator at an appropriate rate and directing the air forced out into the jet of the burner.

THE STANDARD PROCEDURE 153

8.4 PRESENTATION OF THE BEAD

Either the top (curved) or bottom (flat) surface may be presented to the instrument, but, having chosen one or the other, that surface must be used for all beads.

The bead surface to be analysed must be inspected for blemishes, and if any are present a fresh bead needs to be fused. It is normally satisfactory to re-fuse the bead in the same dish provided that the residue of the unpoured fusion is included. Thus, it is best to retain the dish and residue until the bead has been inspected. Re-fusing and re-casting, if correctly carried out, do not seem to affect the results.

If the bottom surface is used, the moulds have to be kept in good condition by frequent polishing, and on any sign of 'bowing' the dish will need to be reformed.

Introduce the beads into the instrument so that the chosen surface is the one to be irradiated by the X-ray tube.

8.5 THE REMAINDER OF THE METHOD

The remainder of the standard method is concerned with instrumental settings, calibration, standards, calculations, etc. These matters are dealt with in other chapters, to which reference should be made.

9 Calibration

9.1 INTRODUCTION

Calibration for XRF is basically the same process as for all physical methods of chemical analysis. It is to establish the response of the instrument to a known amount of the element or oxide sought. In addition to this, any line overlap corrections of one element (or oxide) on the other need to be measured so that they can be allowed for. With most instrumental techniques this is the whole story, as other inter-element corrections usually have to be tackled by matrix matching, complexing the interferents, or reducing the effects by the use of spectrographic buffers etc. In the case of XRF, in principle, the same system is applied; this was usually the case with early work, when determinations were completed by hand computation. Nowadays, however, it is unusual for an instrument, even if a semi-manual type such as the Philips PW 1410, not to be used in connection with a computer, both as a signal recorder and using those signals for the calculations to produce the analysis. Modern instruments such as simultaneous, sequential and combined simultaneous/sequential spectrometers will be computer/microprocessor controlled. They will use the computer to store and process the information produced on the sample as well as for many other functions concerned with the operation of the instrument. In order that this computation can be carried out, all the necessary data concerning the response of a given amount of each of the elements to be determined and the interferences of each of the elements on the others needs to be stored in the computer's memory. Thus, calibration for XRF purposes needs to include, for each oxide to be determined:

 (1) the response of given amounts of the determinand under the chosen operating parameters;

 (2) the response caused, if any, in the absence of the determinand by known amounts of each other component in the sample under the conditions used when measuring the content of determinand (line overlap or line interference) and/or background correction.

 (3) the degree of enhancement or diminution of the signal produced by the determinand caused by the presence of each of the other oxides (inter-element interference, α-corrections).

Thus, calibration for XRF analysis normally has to include the establishment of each of these sets of information. To this has to be added the knowledge of the frequency of the need to correct for any drift in the response of the instrument with time.

In the method described in this book, calibration is undertaken using synthetic standard beads, made by weighing the appropriate amount of each pure oxide, or its equivalent via a suitable compound (usually converted to the oxide on heating), into each bead. It is feasible to use certified reference materials for this purpose but, as has already been discussed, this procedure is subject to several disadvantages. First, there is the fundamental matter of principle that, if certified reference materials are used for calibration, the method is only comparative: using synthetic standards calibration is primary, and the technique is then acceptable for use as a standard method exactly as flame photometry, spectrophotometry etc., which enhances its status. There is also the problem of relying on the certified figures as being absolutely correct: no matter how carefully the standardization has been carried out, errors, even if minor, do occur. These can distort the whole calibration curve.

The response of a given element to the conditions chosen will be dependent not only on those conditions but also on the environment in which it lies. Optimum conditions for exciting and recording the response have been discussed earlier (Chapters 6 and 7). These conditions may conveniently be divided into those basic to the instrument, those that are instrumental (but alterable by the analyst to provide the best parameters for the element sought) and those that are concerned with preparing a bead of maximum value for the type of sample. The first are normally outside the analyst's control, except when selecting the instrument, and the other two usually require some form of compromise between what is ideal and what can actually be achieved. It would, for example, be desirable for chrome ores to be brought into solution in a bead at a dilution of 1:5 so as to improve count-rates and, therefore, precision. Unfortunately, no better dilution than 1:24 (using lithium borate fluxes) can be achieved, and this has to be accepted. Equally, the nature of the sample may mean that an X-ray wavelength of maximum response may have to be rejected owing to the presence in significant amount of an oxide giving rise to unacceptable interference levels. Thus, depending on the number of matrices analysed, more than one calibration may have to be provided for a particular element. In some extreme cases the oxide may have to be determined by alternative means, e.g. soda in the presence of high levels of zinc oxide.

The preparation of a set of calibration curves for a particular type of sample involves a considerable amount of work. Not only is it necessary to prepare synthetic standard beads covering the range of contents for each oxide to be determined, but standard beads will also be needed to determine the levels of interference. Taking the silica/alumina range as an example, the two major content oxides will each need 13 standard binary beads and each minor

CALIBRATION

constituent oxide requires at least three and preferably four beads, solely to establish element response; these are distinct from the beads needed for establishing the levels of inter-element interference, but effectively include those for line overlap effects, as they can be used for this purpose. After the set of calibration curves for the first type of sample has been prepared, it is possible to use some, at least, of the same beads for other types of samples, provided that the same flux and the same flux/sample ratio are used.

9.2 STANDARDS

Standard materials are usually oxides, although this is not always possible. Many oxides (e.g. Na_2O) are not stable to atmosphere, in which case it is necessary either to apply prior treatment before weighing out the requisite mass or to use other and more stable compounds of the same element. Even where oxides may be used with success, it is almost inevitable that some form of 'drying' procedure will be needed to ensure that the correct weight of oxide is incorporated in the bead. As already discussed, almost all powders will adsorb moisture on to the particle surfaces. With some compounds this effect may be small but others, not necessarily thought of as hygroscopic, will adsorb considerable amounts. Even small adsorptions can be significant for major contents. In any event, the mode of correction of these sorts of possible error is usually so easy that 'drying' should be regarded as routine.

Some 'metallic' elements have oxides that are unsuitable as weighing forms, as are many acidic oxides. Typical examples are the alkali oxides on the one hand and sulphur trioxide on the other. In almost all cases, where oxides cannot be used, it is possible to use the carbonates as a weighing form for the 'metallic' elements and lithium salts for the acidic oxides or elements. 'Drying' of the weighing form is still essential.

Although carbonates such as sodium carbonate do not normally decompose on heating, it is safe to accept that after melting with other constituents in a bead the carbon dioxide has been lost. This even applies to standard beads simulating basic materials such as lime or magnesia, since the borate in the flux will cause evolution of the carbon dioxide. When using the lithium salt of an anionic element or oxide, e.g. fluoride or sulphate, the flux needs to be modified to allow for the equivalent amount of lithium being added. The Li_2O content of the reagent, correctly in the case of lithium sulphate and similar salts or its calculated equivalent in the case of, say, lithium fluoride or chloride, needs to be regarded as part of the flux. This normally will entail reducing the weight of flux and adding boric oxide or modifying the ratio of meta- to tetra-borate.

The materials selected as standards need to be in a high state of purity, at least with reference to the presence of other determinable consituents. This is important, not only for ensuring that the true intended weight of oxide is present in the standard bead, but also to ensure that no significant amount of any other oxide to be determined is added. Generally speaking, it is necessary to obtain reagents of spectrographic purity grade (e.g. Johnson–Matthey 'Specpure'), especially when the oxide is to be used as a major content addition. Occasionally it is possible to obtain supplies of some oxides in pure form from other sources much more cheaply, so that it is as well to investigate any such possibilities, always ensuring that the level of purity is as claimed. Most oxides will be in the form stated, e.g. silica will be, for all practical purposes, SiO_2, and will not contain identifiable amounts of sub-oxides. On the other hand, a number of oxides of elements with variable valency states cannot be relied on to be totally in the correct state, or even totally in the form stated in the supplier's catalogue. Neither should it be assumed that, just because an oxide can be warranted not to contain contaminant impurities, that it is in a form, or even can easily be converted into a form, that can be weighed out to give a prescribed concentration in a standard bead. These types of oxide may even necessitate some form of assay by other than XRF means to ensure their correct concentration in the material as supplied and pre-treated. However, since this will destroy the primary nature of the XRF technique, apart from loss of accuracy, all practical means should be taken to avoid it.

For the various reasons given, virtually all the oxides or other compounds require some form of pre-treatment, generally drying or pre-heating to specified temperatures. Table 9.1 gives details of appropriate pre-treatments for most of the oxides or elements discussed in this book. The table also includes a number of additional notes about specific oxides. For example, silica can be obtained in a number of physical forms, precipitated from solution, hydrolysed from (say) silicon tetrachloride, fused and ground or from pure natural quartz. Experience has shown that very powdery silica is difficult to handle and adsorbs moisture both very quickly and inhomogeneously, making it a poor standard material. Fused silica is much better. In fact, even 'Specpure' silica has been delivered in different physical forms, some granular and some not. Some discussion with suppliers or selection of sources of supply is advisable, particularly with those materials likely to be used in significant amounts. Considerable sums of money can be saved by buying in sensibly sized batches and from selected sources of supply.

9.3 CALIBRATION

9.3.1 Establishing Calibration Curves

It is imperative that an accurate definition of the calibration curves and inter-element effects is achieved. Extra time and care spent at this stage means greater certainty in the subsequent analyses. It is certainly worthwhile to increase counting times at this stage to improve precisions.

CALIBRATION

Table 9.1. Materials used for standard beads in alphabetical order of formula. Factors quoted are to convert the weight of oxide to compound

Oxide etc.	Compound	Heat treatment	Comments
Al_2O_3	Al_2O_3	1200°C	Should be in granular form
B_2O_3	B_2O_3	700°C	Weigh out corrected for 700°C loss
BaO	$BaCO_3$	220°C	Factor 1.3913
CaO	$CaCO_3$	220°C	Factor 1.7848
CoO	Co_3O_4	700°C	Factor 1.0711
CuO	CuO	700°C	
Cr_2O_3	Cr_2O_3	1000°C	
F	LiF	110°C	Factor 1.3653[a]
Fe_2O_3	Fe_2O_3	700°C	Can use temperatures 700–1000°C
HfO_2	HfO_2	1000°C	
K_2O	K_2CO_3	220°C	Factor 1.4672
Li_2O	Li_2CO_3	220°C	Factor 2.4729
MgO	MgO	1200°C	
Mn_3O_4	MnO_2	1000°C	Factor 1.1399
MoO_3	MoO_3	700°C	
Na_2O	Na_2CO_3	220°C	Factor 1.7101
Nb_2O_5	Nb_2O_5	1025°C	
NiO	NiO	1025°C	
P_2O_5	$LiPO_3$ or	700°C	Factor 1.2105[a]
	$Ca_2P_2O_7$	700°C	Factor 1.7986
PbO	PbO	500°C	
SO_3	Li_2SO_4	220°C	Factor 1.3732[a]
Sb_2O_3	Sb_2O_3 or	220°C	
	Sb_2O_4	220°C	Factor 1.0549
SiO_2	SiO_2	1200°C	Should be in granular form
SrO	$SrCO_3$	220°C	Factor 1.5791
TiO_2	TiO_2	1000°C	
V_2O_5	V_2O_5	220°C	
WO_3	WO_3	110°C	
Y_2O_3	Y_2O_3	1025°C	
ZrO_2	ZrO_2	1000°C	

[a]The equivalent content of lithia must be allowed for in the flux composition.

As the ultimate accuracy of the results depends on establishing the correct response curve, the greatest of care needs to be taken both in the preparation of standard beads and in their subsequent presentation to the instrument. Operating parameters should have been selected to give optimum response bearing in mind the possibilities of line overlap interference etc. Ideally one would always use the most sensitive line (giving attention to different crystals and wavelengths), but this choice may well be restricted by inescapable interferences; another set of conditions may give better selectivity, albeit with poorer limits of detection.

Thought also needs to be given, at the outset, to the level of accuracy and precision demanded of the analysis. It is not unknown for these to be more stringent than are actually obtainable. XRF, like any other analytical technique, has 'natural' levels of accuracy and precision that arise using the technique in a sensible and economically viable fashion. When such difficult demands are made it is essential that the precise requirements are discussed with the 'client'. This needs if possible to lead to compromise. It may demand, perhaps, a more realistic specification from the ultimate user or, alternatively, an improvement in technique if practicable (and economically feasible), or the use of other analytical techniques if this will better serve the purpose.

Major content determinations require, in general, a high level of accuracy. To obtain precisions of $\pm 0.1\%$ (2σ) absolute at the 50% level of content requires relative precisions of 0.2%. To achieve this, count numbers need to be high, so that for a given count-rate a long exposure time may be necessary. As many applications described here are concerned with relatively light elements, such as silicon and aluminium, counting times of 100 s are not thought excessive. The advent of InSb crystals has reduced this time for silica, but alumina still needs a long counting time. Using a simultaneous instrument the counting time for all can be the same as that for the longest. With sequential instruments offering the ability to select times for each oxide, times will obviously be chosen so as to use the minimum time consistent with each oxide's requirements.

9.3.2 Composition of Standard Beads for Calibration Curves

Although the same general principles apply to all types of materials, it is convenient to illustrate them with the most common, the silica/alumina range. This example has the additional advantage that it contains two major contents and serves to illustrate how that situation is treated. Obviously, where there is only one major constituent and the remaining components are in the minor content range the position is simpler. In fact, in that situation it may well be that for some purposes inter-element effects can be ignored. Generally, however, if the establishment of inter-element effects is not too time-consuming, it is usually worth doing, as results will by definition be that much more accurate.

It is usual to select the major constituent of the sample type as the base from which to work. Thus, a zero-impurity standard sample bead would contain 100% of the chosen oxide in the appropriate weight of flux. In the case of the silica/alumina range this oxide would normally be silica, but could equally well be alumina. Calibrations for the other oxides to be determined are achieved by replacing the appropriate weight of silica with the required weight of the oxide in question. For the calibration for silica, alumina needs to be used as the base oxide.

The thirteen binary standards for the calibration of silica and alumina are shown in Table 9.2.

The standard beads are prepared exactly as described for the preparation of sample beads; the oxides and the flux are used in the 'dry' (ignited and cooled)

CALIBRATION

Table 9.2. Composition of standard beads for Al_2O_3 calibration silica/alumina range materials: for a 9 g bead

Percentages		Weights		
Al_2O_3 (%)	SiO_2 (%)	Al_2O_3 (g)	SiO_2 (g)	1/4 Flux (g)
0	100	0.000	1.500	7.500
5	95	0.075	1.425	7.500
10	90	0.150	1.350	7.500
20	80	0.300	1.200	7.500
30	70	0.450	1.050	7.500
40	60	0.600	0.900	7.500
50	50	0.750	0.750	7.500
60	40	0.900	0.600	7.500
70	30	1.050	0.450	7.500
80	20	1.200	0.300	7.500
90	10	1.350	0.150	7.500
95	5	1.425	0.075	7.500
100	0	1.500	0.000	7.500

state, using a total weight of oxides of 1.500 g, i.e. simulating a sample. As discussed elsewhere, it is acceptable to use the equivalent weights of undried materials if these are known.

Where sufficient numbers of either high-silica or high-alumina materials are to be analysed, it may be desirable to prepare separate calibrations over the shorter ranges (i.e. alumina or silica respectively) using four or five points to cover the range, say 0–5% or 0–10%. This should improve the precision with which the appropriate portion of the calibration curves will be defined, leading in turn to more accurate results. Some analysts combine the results from the short and long range calibrations, arguing that this gives more weight to the lower end of a single calibration, and thereby avoiding separate calibrations. There is a danger that this technique, giving equal weight to each point, will distort the longer range calibration. Various attempts have been made to apply differing 'weights' to the points at the lower end of the range, but this can only be empirical. Such devices are best avoided. Results of a higher degree of absolute accuracy are clearly needed for small amounts of silica in alumina, and vice versa, than are tolerated in the middle of the range. Even when extraordinary care is taken, results at the 50% level for either oxide cannot be guaranteed to be better than 0.2% (for ordinary work $\pm 0.5\%$ would normally be claimed); this is totally unacceptable at fractional percentages.

For minor consituents, up to 5% content, it would be usual to establish four calibration points, roughly equally spaced. Thus, contents of the oxide being calibrated would be, say 0%, 1.7%, 3.4% and 5%, with silica making up the residue to provide an apparent 100% total 'sample', i.e. 1.5 g of SiO_2 minus the weight of oxide being calibrated.

9.3.3 Checking the Calibration Curves

One of the important advantages of using synthetic standards is that there is no doubt that the points on the concentration axis are correct, whereas with analysed standards some doubt, however small, remains. In addition, standards are only binary, so that complications due to line interference and α-corrections are either absent or, at worst, easily corrected for. It is possible, in fact advisable, to plot the results on graph paper. Most calibrations will be straight lines, a few will be smooth curves. Either way, visual inspection of the graphs will show any point or points not dropping on the line. This enables repeat runs on the instrument, or allows new beads to be prepared, as necessary, to check for possible errors. Normally this will resolve the problem (most such errors appear to be operator error, usually in weighing); in the unlikely event that the 'sport' result is confirmed it becomes clear that some explanation has to be sought.

The use of analysed samples containing all the constituents will produce counts reflecting the interferences (line overlap and α-corrections) as well as the oxide response sought. This problem can only be resolved by the use of multiple regression analysis. Because the composition of CRMs is beyond the control of the analyst, their spread of individual oxide contents is unlikely to be that best suited to show up both calibration shape and interference effects. It is not difficult to visualize circumstances where this could lead to erroneous calibrations and interference coefficients. Multi-regression analysis applied to such beads could fail to identify analyst errors, leading to disastrous results. On the other hand, the range of oxides in multi-component synthetic standards is notoriously difficult to weigh out accurately and without human error. The latter danger can be reduced by a dilution technique, i.e. by carefully making up beads with an appropriate mix of oxides and then grinding. After this, a weight of the ground material equivalent to the weight of oxide(s) required is taken. Alternatively, aqueous solutions can be prepared at moderate dilutions and appropriate (small) volumes added as required to the flux mix. After drying the mix can be fused. In any case, multi-element standards are best made up in duplicate. If their intensities do not check then repeats should be made until the true intensities have been defined.

Another practice not infrequently met with is that of using the same chemically analysed samples both for calibration and as unknowns, but this is unacceptable. The latter use is intended either to check the calibration or prove the capabilities of the method advocated. Checking calibrations by using the same samples is self-defeating, as the same counts will be recorded on the sample as on the standard. This will lead to apparently correct answers, as the true counts from the determinand will have been modified by precisely the same interferences in both cases. Attempts to analyse other samples of the same type but with different compositions may, and often do, yield incorrect results because interference effects have not been properly identified. It is clear, therefore, that achievement of correct answers for 'samples' consisting of standard reference materials that have been used in the calibration does not necessarily show that the calibrations are correct. This difficulty can sometimes be alleviated by using mixtures of certified reference materials as 'samples', thus achieving contents deviating from the calibration points and containing levels of impurities different from those during calibration. Even this approach, however, is best avoided.

The fused, cast bead technique allows standards to be made as desired, and provides the simple binary or ternary standards which are ideal for calibration. CRMs or other accurately analysed chemical standards, on the other hand, have then a vital role to play in testing the validity of the calibrations and in the training, or monitoring the performance, of fledgling analysts. In this respect the analyst of oxide and similar materials is in a much better situation than the metal analyst, who has no alternative but to use multi-regression techniques.

9.3.4 Presentation of the Standard Beads to the Instrument

As discussed earlier, in order to eliminate the possible effect of medium and long term drifts in the response of the instrument, all readings are made as ratios to the response of a set of standard beads. These beads include a zero point for all elements as well as a content near to the top point of each calibration, i.e. giving a theoretical ratio of $\simeq 1$ for the top standard. In principle it would be possible to use any material giving an adequate response under the parameters used, so long as the identical material was used throughout the life of the calibration. These standards are really sensitivity ratio standards but are generally simply known as ratio standards (re-calibration standards or point-monitor standards). It is also advisable to monitor the zero response of each oxide, running a zero standard, to check for any significant change in background effects from previous calibrations, or for contamination of the zero ratio standards in use.

In addition, checks need to be made when a new batch of flux is brought into use. Provided that the reagents used are of good quality this should not provide much difficulty, but experience has shown the advisability of maintaining this check. Change of flux batch occasionally produces small changes in zero readings because of slight alterations in impurity levels. These variations are, of course, made more significant because five times as much flux is used as sample, so that the equivalent of 0.01% of an oxide impurity would register as 0.05% content. Trace elements that have been found to vary in content from one batch of flux to another include Si, Al, Mg, P and S. On initial calibration, however, these zero readings more or less have to be accepted unless there is any reason to suspect the results, suggesting some unacceptable degree of contamination. On the other hand, although theoretically possible, calibration slope factors do not appear to be affected by a change in the batch of flux.

In the example being used here, viz the silica/alumina range, the ratio and zero standard beads are as follows.

The 100% silica standard bead serves as the ratio standard for silica, and also as the zero, or background, standard bead for alumina, titania, ferric oxide, lime, magnesia, soda, potash and tungstic oxide etc.

The 100% alumina standard bead serves as the ratio standard for alumina and the zero standard for silica.

A third standard bead is needed to compensate for any possible changes in sensitivity for the minor constituents. A suitable composition for this is:

$$SiO_2 = 63\% \quad Na_2O = 10\% \quad Fe_2O_3 = 5\%$$
$$Al_2O_3 = 0\% \quad WO_3 = 2\% \quad MgO = 5\%$$
$$CaO = 5\% \quad TiO_2 = 5\% \quad K_2O = 5\%$$

The calibration curve for each oxide is established by running each of the standard beads for the oxide on the instrument, at the appropriate parameters, and recording the counts. These counts are converted to ratios, R, as shown below. It is best to make up and run a spare set of ratio standards in case of loss or damage.

For silica

$$R_{Si} = \frac{N_S - N_A}{N_{100S} - N_A}$$

where R_{Si} is the ratio for the particular standard point being run, N_S is the count for silica obtained on the bead being run, N_A is the count for silica on the 100% alumina bead, N_{100S} is the count for silica on the 100% silica standard bead.

For alumina

$$R_{Al} = \frac{N_A - N_S}{N_{100A} - N_S}$$

where R_{Al} is the ratio for the particular standard point being run, N_A is the count for alumina obtained on the bead being run, N_S is the count for alumina on the 100% silica standard bead, N_{100A} is the count for alumina on the 100% alumina standard bead.

CALIBRATION

For the minor constituents

$$R_M = \frac{N_M - N_S}{N_{TM} - N_S}$$

where R_M is the ratio for the particular standard point being run, N_M is the count for the appropriate minor constituent oxide obtained on the bead being run, N_S is the count for the appropriate minor constituent oxide on the 100% silica standard bead, N_{TM} is the count for the appropriate minor constituent oxide on the minor constituent ratio standard bead.

Once calculated, the ratios are used for plotting and/or calculating the equations for the calibrations, which in turn are usually stored on a computer and are used ultimately for calculating the analysis of samples.

An acceptable alternative to the calculations given above is to correct back to the count-rates obtained when the ratio standards were first run alongside the calibration standards. This procedure is given as an Appendix to BS 1902: Part 9.1.

9.3.5 Determination of Line Interferences

The top calibration standards for each oxide are used for establishing the levels of line interferences on other oxides. With the instrumental parameters set to be suitable for, and as normally used for, the oxide being determined, the top standard bead for each of the other oxides is run to establish the response, if any, apparently for the oxide being determined. The apparent content of determinand caused by the presence of each 1% of the interfering oxide is calculated for subsequent use in analytical calculations and normally stored in the computer. The effect can be negative where it is due, not to a line overlap, but to a decrease in background intensity. Normally, unless trace levels are being sought, off-peak backgrounds are not monitored. It is sometimes useful to run other levels of interfering elements in order to check that the apparent interference is not due to contamination of the bead (e.g. by sodium on magnesium), and to confirm that the interference is linear. As an example, using a rhodium tube, the effect of zirconia on soda and magnesia appears to be quadratic.

For example, if the 100% Al_2O_3 standard bead yields a figure apparently equivalent to the presence of 0.16% of MgO under the parameters used for the determination of magnesia, then the line interference correction (I) of Al_2O_3 for MgO is $0.16/100 = 0.0016$.

9.3.6 Determination of Inter-element Effects (α-Coefficients)

There is, in general, no need to establish the effect of silica (the matrix oxide) on either alumina or the minor constituent oxides, as the effect is small for alumina and the effect of variations in silica content can be neglected for the relatively short range calibrations of other constituents. In the latter case corrections do not contribute to overall accuracy. It is necessary, in general, to establish the effect of the other oxides on each other. Theoretically all these determinations need to be carried out, but in practice this is not so. Some effects are so small that little is to be gained by their applications, and in other cases the content of interfering oxide is so small that the net effect on the determinand is negligible, e.g. the effect of tungstic oxide (from the grinding vial) on other constituents, since the content of tungsten is so low. Calculation of theoretical mass absorption corrections so as to estimate the need for actual determination in the material type in question is also possible. In many cases, if the model is sophisticated enough, the use of theoretically calculated corrections can give an adequate amelioration of the error without recourse to binary standards. Even with a good mathematical model, it is advisable to check large α-coefficients experimentally, as well as those mass-absorption effects close to an absorption edge.

Inter-element corrections of this type are usually called 'α-corrections' to distinguish them from line interference corrections (line overlaps). An α-correction is the figure arrived at by applying the appropriate α-coefficient to the respective contents of interferent and oxide being interfered with. In most cases a theoretical correction programme is not available, and standard beads need to be made to establish α-coefficients. These beads can be split into three groups. First, there are those for establishing the effect of minor constituents on alumina and vice versa. Second, the standard beads used for the calibrations are used to determine the effect of these oxides on silica. Last, there are the beads necessary for the determination of the effects of minor constituents on each other. The compositions of the first and third groups of beads needing to be made are shown in Tables 9.3 and 9.4, together with the uses for each bead.

It is as well to check the experimental results obtained for the various α-corrections against specimen values, a few of which are shown in Tables 9.5

Table 9.3. α-Correction beads for minor constituents in alumina[a]

%	Standard bead Oxide	% Al_2O_3	α-Corrections established
5	TiO_2	95	TiO_2 on Al_2O_3 : Al_2O_3 on TiO_2
5	Fe_2O_3	95	Fe_2O_3 on Al_2O_3 : Al_2O_3 on Fe_2O_3
5	CaO	95	CaO on Al_2O_3 : Al_2O_3 on CaO
5	MgO	95	MgO on Al_2O_3 : Al_2O_3 on MgO
5	K_2O	95	K_2O on Al_2O_3 : Al_2O_3 on K_2O
10	Na_2O	90	Na_2O on Al_2O_3 : Al_2O_3 on Na_2O

[a]These standards also yield line interference or background effects on silica (Si $K\alpha$ line) for the appropriate minor constituents

Table 9.4. α-Correction beads for minor constituents on each other

%	Standard bead 1st oxide	%	2nd oxide	% SiO₂	α-Corrections established
*5	TiO₂	5	Fe₂O₃	90	TiO₂ on Fe₂O₃ : Fe₂O₃ on TiO₂
*5	TiO₂	5	CaO	90	TiO₂ on CaO : CaO on TiO₂
5	TiO₂	5	MgO	90	TiO₂ on MgO : MgO on TiO₂
*5	TiO₂	5	K₂O	90	TiO₂ on K₂O : K₂O on TiO₂
5	TiO₂	10	Na₂O	85	TiO₂ on Na₂O : Na₂O on TiO₂
*5	Fe₂O₃	5	CaO	90	Fe₂O₃ on CaO : CaO on Fe₂O₃
*5	Fe₂O₃	5	MgO	90	Fe₂O₃ on MgO : MgO on Fe₂O₃
*5	Fe₂O₃	5	K₂O	90	Fe₂O₃ on K₂O : K₂O on Fe₂O₃
*5	Fe₂O₃	10	Na₂O	85	Fe₂O₃ on Na₂O : Na₂O on Fe₂O₃
5	CaO	5	MgO	90	CaO on MgO : MgO on CaO
*5	CaO	5	K₂O	90	CaO on K₂O : K₂O on CaO
5	CaO	10	Na₂O	85	CaO on Na₂O : Na₂O on CaO
*5	MgO	5	K₂O	90	MgO on K₂O : K₂O on MgO
*5	MgO	10	Na₂O	85	MgO on Na₂O : Na₂O on MgO
5	K₂O	10	Na₂O	85	K₂O on Na₂O : Na₂O on K₂O

The number of standards shown in the table may be reduced to the ten marked with an asterisk, using calculated results for the remainder.

Table 9.5. Typical α-coefficients for chromium side-window tubes. All results × 10⁵

Oxide		SiO₂	TiO₂	Al₂O₃	Fe₂O₃	CaO	MgO	K₂O	Na₂O
Line									
Si	Kα	0	200	356	284	124	282	59	230
Ti	Kα	0	0	−41	−139	970	−92	856	−107
Al	Kα	0	109	0	335	100	337	37	293
Fe	Kα	0	1160	−60	0	1059	96	1105	−88
Ca	Kα	0	−46	−38	−55	0	−65	1033	−104
Mg	Kα	0	100	−49	284	68	0	47	340
K	Kα	0	−66	−14	−11	−177	−114	0	−132
Na	Kα	0	0	−92	274	112	−368	−5	0

Using the PW 14 series instruments; incident angle 63°, take-off angle 40°.

and 9.6. The α-coefficient determinations can be prone to large errors due to slight defects in the bead, etc., so that more care is needed here than during the actual process of establishing calibration curves. This, of course, is most important when the level of correction is high, either because the factor is large or the amount of interfering oxide results in large corrections. When using sequential spectrometers, counting times per channel are generally shorter than with a simultaneous instrument, thus tending to yield poorer counting statistics. It is therefore best to increase counting times for sequential spectrometers when α-coefficient data is being collected.

Table 9.6. Typical α-coefficients for rhodium end-window tubes. All results × 10^5

Oxide		SiO$_2$	TiO$_2$	Al$_2$O$_3$	Fe$_2$O$_3$	CaO	MgO	K$_2$O	Na$_2$O
Line									
Si	Kα	0	72	290	245	54	259	54	216
Ti	Kα	0	0	−49	−169	1043	−146	985	−125
Al	Kα	0	106	0	272	65	280	43	249
Fe	Kα	0	1060	−57	0	1096	−151	1036	−173
Ca	Kα	0	−279	−45	−176	0	−74	1001	−121
Mg	Kα	0	193	−11	301	54	0	0	103
K	Kα	0	−300	−70	−174	−248	−188	0	−128
Na	Kα	0	138	−15	298	108	0	112	0

Using a Telsec TXRF, incident angle 90°; take-off angle 38.8° for Al and Si, 37.2° for Na and Mg, 34.8° for Fe, 32.5° for Ti and 26.5° for K and Ca.

9.4 NOTES ON CALIBRATION

(1) *For all the oxides above, except silica and alumina.* Calibrations should be straight lines for the ranges discussed. No accuracy is gained in correcting for the inter-element effects of silica on the other oxides, and such a correction involves considerable extra effort.

(2) *For alumina.* The inter-element coefficient of silica on the aluminium Kα line is small, and no significant accuracy is gained from further calculations. The calculation should normally be as for a straight line, but it may be necessary to express it as a quadratic (or even a cubic) equation with low non-linear terms in order to obtain the best fit. High non-linear terms are usually indicative of a spectrometer problem (e.g. dead time effects) that needs investigation and eradication.

(3) *For the silicon Kα line.* The inter-element effect due to alumina is large and a correction needs to be made. Of the many possible ways of calculating the inter-element effect of the matrix oxide, the following has been found to be both simple and effective.

A calibration ratio is calculated from the result from each of the silica and alumina binary standard beads. A factor E is calculated by dividing the ratio into the concentration of silica, and is expressed as a percentage. Each E is then plotted on the y axis on graph paper against the interfering oxide concentration (in this case Al$_2$O$_3$), again expressed as a percentage, on the x axis; this graph should be a straight line sloping from bottom left to top right. The intercept on the y axis gives E_0, that is, the slope of the calibration for silica with no alumina present. The slope of the graph K (a positive sign being indicated by an increase from left to right) is calculated as E per 1% of Al$_2$O$_3$. The α-correction is defined as K/E_0. The corrected ratios (R_T) now need to be produced from the ratios determined directly from the standard beads (R_A) by using the equation:

CALIBRATION

$$R_T = R_A(1 + \alpha C)$$

where α is as derived above (K/E_0), and C is the concentration of Al_2O_3 in the standard bead.

The calibration is now converted to the relationship of concentration (%) of the oxide against the values of R_1, and should be linear. These values of R_T are used in the calculation of α-coefficients of other oxides on SiO_2.

Any standard bead producing a result that lies further off the calibration curve than three times the standard deviation calculated from counting statistics needs to be checked, first by being re-run and then, if necessary, by fusing a further bead to replace that point of the calibration.

9.4.1 Determination of Other α-Coefficients

It is necessary to derive first the true ratios (R_T) from the established calibration graphs. The inter-element effects are measured using the beads prepared for this purpose, and are exemplified in Tables 9.5 and 9.6. The ratios derived from the α-coefficient standard beads need first to be corrected for any background or line interference effects, so that the response used is that deriving solely from the determinand. This gives a value for each oxide of the apparent ratio (R_A). The α-coefficient is then calculated by using the equation:

$$\alpha = \frac{R_T - R_A}{C_J . R_A}$$

where C_J is the concentration (as a percentage) of the interfering oxide.

As an example, a standard bead containing 5% of CaO and 5% of TiO_2 gives a recorded ratio of 0.9492 for TiO_2. Correcting this for a line interference derived from running a 5% CaO/95% SiO_2 bead (giving a ratio of 0.0005), the ratio becomes 0.9487. A standard containing 5% of TiO_2/95% of SiO_2 gives a true ratio (R_T) of 0.998. Therefore, using the above equation, the α-correction for lime on titania becomes:

$$\alpha = \frac{0.998 - 0.9487}{0.9487 \times 5} = 0.01039$$

9.5 SYSTEMATIZING CALIBRATIONS AND FUSIONS

9.5.1 Introduction

Once the first base calibration has been prepared and is in use, it will soon be realized that there are major advantages in systematizing calibrations and fusion procedures. This section suggests how this can be achieved.

9.5.2 Unifying Calibration and Fusion Systems

Although it is possible that a few laboratories only need to analyse a single type of material, necessitating only short range calibration curves, most will be called on to analyse a variety of materials. These will usually include some in the silica/alumina range together with one or more other types. The first calibration made will be that most commonly used; it is convenient for the purposes of this discussion to assume that this will be that for material(s) in the silica/alumina range. Setting up calibrations for the material(s) will involve preparing ratio standards, establishing calibration curves, generating α-coefficients and measuring line overlap or background corrections. Although this will involve a great deal of work, time and thought, it will already have been accepted as being worthwhile, in that an XRF spectrometer has been purchased.

Once the original calibration has been set up (and is presumably in use), attention will be turned to the next class of material to be covered. Three possibilities exist:
(1) a different flux may be required;
(2) the same flux may be used but a different sample/flux ratio;
(3) flux conditions may be the same.

If there is a choice of which type of material to tackle next it is logical to attempt one which utilizes conditions as near identical to those already in use as possible. If, for example, the first calibration was for an aluminosilicate and it is desired to analyse talc, this will entail a relatively small amount of effort. The magnesia calibration would require extension, and a few α-coefficient standards would be needed to better define the effects on and of magnesia, together with making a magnesia ratio standard bead. With these additions and 90% of the aluminosilicate calibration system the laboratory becomes equipped to analyse talc. Similarly, a long range calibration for lime, etc., extends the aluminosilicate calibration to cover cements, calcium silicates and aluminates.

This approach soon creates a sense of unification in the structure of the calibration system, and helps the analyst to appreciate the ease with which new calibrations can be added for different types of materials using minimum effort. In the groups so set up there may be requirements for more closely defining a minor content (e.g. Al_2O_3 in high-silica materials) or similarly defining a short range at an intermediate point in the calibration (e.g. Al_2O_3 in cements). This would require, in either case, a set of standard beads for the oxide in question more closely spaced to cover the desired range.

Other benefits accrue from this technique. Re-calibration after changing a batch of flux entails less effort. As compositions are defined round silica (or alumina) rather than the average composition of the material type in question, addition of a new element is much simpler. The same sort of argument applies to the situation where the matrix is based on other elements, e.g. calcium for limestones and dolomites. Instrumental checks are more easily made and faults are revealed more readily. The latter benefit is less obvious in principle, but it proves so in practice.

This was the system adopted by the authors from the outset, and a whole host of calibrations has been generated from the original silica/alumina one. Other series of calibrations arose naturally from the need to use other matrices, other fluxes and different flux/sample ratios. The next section outlines some of the possible sets and itemizes some of the modifications needed to cope with individual sample types.

9.5.3 Some Unification Series

9.5.3.1 Base Calibration—Silica/alumina: 4:1 flux/sample ratio 5:1

This calibration covers a wide range of compositions from siliceous to aluminous materials. The low melting point and low acidity of the flux help in retaining some volatiles such as sulphur trioxide. The scope is extended by the ability to use lower fusion temperatures (1050°C) where necessary or desirable. Some types of sample when dissolved in this flux and at this dilution may tend to devitrify on casting, but these are relatively few among ceramic and geological samples. They are usually easily identified by their inherent basicity or by their major oxide's known poor glass-making properties.

If the flux is to be used for *high-silica materials*, an additional short range calibration is needed for Al_2O_3 at the lower end of the range so as to establish alumina figures more accurately.

Similarly, if the calibration is to be used for *high-alumina materials*, an additional calibration is needed for SiO_2 at the lower end of the range.

Cement and calcium silicate require, in addition to the normal calibration, a medium range calibration for alumina at suitable levels of content, a long range calcium calibration and a calibration for sulphur trioxide.

Bauxite requires an additional long range calibration curve for Fe_2O_3 if 'raw' bauxites are to be attempted as well as refractory bauxites.

Slag (from ferrous metal production) utilizes the modified calibration as for calcium silicates, and would also involve the use of the long range ferric oxide calibration (possibly that of raw bauxites) and possibly some other oxides such as manganese, etc.

Fluorspar involves the calcium silicate calibration, but inter-element corrections would need to be applied for the fluorine.

Talc, steatite and soapstone require an extended magnesia calibration and an intermediate level Al_2O_3 calibration.

Bone china body and bone ash involve the use of the calcium silicate calibration together with a long range P_2O_5 calibration, with the addition of inter-element corrections for and on phosphorus pentoxide. The calibration curve for P_2O_5 needs to be long enough to cover bone ash and, for the latter, the silica calibration for high-alumina materials would be used to cover the low levels of SiO_2.

Zircon utilizes the same calibrations as for high-silica materials, but with appropriate range calibrations for ZrO_2 and HfO_2 together with the necessary inter-element corrections.

Glazes, glass and lead borosilicate utilize aluminosilicate calibrations. Additional calibrations and corrections need to be made for any additional oxides, e.g. PbO, ZnO, SnO_2 etc. In addition, corrections have to be applied for the boric oxide contents.

9.5.3.2 Base Calibration—Limestone: flux lithium tetraborate; flux to sample ratio 5:1

This flux is more acidic and therefore better for fusing major contents of basic oxides. It has the disadvantages of having a higher melting point and being very viscous even at a high temperature (1200°C).

Dolomite needs an extended magnesia calibration.

Cement and calcium silicate may be handled using this calibration by extending the silica and alumina calibrations and adding a calibration for sulphur trioxide. This is an alternative to the use of the silica/alumina calibration. However, the possibility of retaining sulphur trioxide in a lithium tetraborate flux will be much lower.

Bone china body and bone ash could utilize calibration as for cements, with the addition of a long range calibration for P_2O_5 and corrections for and on phosphorus pentoxide, as an alternative to the use of the silica/alumina calibration.

Fluorspar requires the application of corrections for the fluorine remaining in the bead. Melt weights are required to identify this loss during fusion, as it is greater in this flux than in the 4:1 flux. This is an alternative to the use of the silica/alumina calibration.

Silica/alumina range materials can be handled using the base limestone calibration. This may be useful if the main work of the laboratory is the analysis of limestone and calcium silicates, e.g. that at a cement works. The calibration used for the analysis of cement can be applied. This course would involve the preparation of long range calibrations for SiO_2 and Al_2O_3 and the extension of some of the minor constituent calibrations. This is an alternative to the use of the silica/alumina calibration.

9.5.3.3 Base Calibration—High Magnesia: flux lithium tetraborate; flux/sample ratio 10:1

Magnesites, etc., need a greater dilution, even in the acidic tetraborate flux.

CALIBRATION

If magnesites etc. are the main types of samples analysed and the laboratory's base calibration reflects this, it is possible to extend the calibration to provide a set of unified calibrations on the same principles as above. A long range calibration for CaO, for example, would cover the analysis of dolomites, limestones and cements. Other classes of material could be handled successfully by similar additional calibrations, provided that the increased dilution causes no problems with lack of sensitivity.

The same flux handles a number of oxides such as those of iron and manganese, producing excellent beads.

9.5.3.4 Base Calibration—Zirconia: flux lithium tetraborate; flux/sample ratio 12:1

If zirconia is the most analysed sample, demanding an appropriate base calibration, this can be extended to cover zircons by the addition of a long range SiO_2 calibration. Appropriate additional calibrations and, possibly, inter-element corrections allow the analysis of zirconates.

Titania and titanates fuse well in this flux at this dilution, so that the calibration for zirconia could have factors in common with these sorts of materials.

9.5.4 Alternative Approaches to Material Types

From Section 9.5.3 it will be seen that many materials can be handled using different parameters. Cements, for example, may be analysed using 4:1 flux at a dilution of 5:1, or in lithium tetraborate at the same dilution. Similarly, zircons, bone ashes, bone china bodies and dolomites can be fused and analysed in two different ways. In this book, parameters have been recommended that have proved eminently satisfactory for the materials in question. Nevertheless, accurate results can be achieved with a variety of fluxes and flux/sample ratios. In principle, if a type of material can be fused in a particular flux at a given flux/sample ratio then it could be calibrated for and analysed using calibrations common to those used for the base material in the set, with the addition of the necessary additional calibrations etc. It is even possible to use greater dilutions than necessary provided that this entails no difficulties with sensitivities. Problems can arise from volatilization or devitrification; if so the particular parameters may prove unacceptable.

9.5.5 Standard Methods—National and International

Within a single laboratory, it is of little consequence which base calibration is chosen provided that the results are of the required standard. However, if it is necessary to compare results with those from other laboratories to provide referee quality analyses, it is a major advantage to be able to claim to have used a standard procedure. A multiplicity of variants on the same basic method gives rise to great difficulties in agreeing a single National Standard and even more so an International Standard Method.

National Standard Methods are developed by organizations such as BSI, ASTM, DIN (Deutsche Institut für Normung), SFC (Société Francaise Céramique), etc., while international work is handled by ISO and, for Europe, CEN (Committée Européen de Normalisation) or previously, for the refractories industry itself, PRE (Federation Européen des Fabricants de Produits refractaires). Even nationally it is sometimes difficult to achieve total agreement to a single variation; this is seen in the casting procedures of the BSI methods in BS1902. Internationally, where countries have already accepted different approaches, it is often almost impossible to get agreement on the optimum parameters. Even basic philosophies may be at variance, so that acceptable compromises that yield essential simplicities or accuracies can be very difficult.

At the time of writing, with the very important year for Europe (1992) already upon us, a great deal of effort is having to be directed to harmonizing BS, CEN and ISO methods. This is an area of activity to which Ceram Research is heavily committed. Dr Oliver sits on the four National committees (BSI) dealing with bauxite, glass, cements and iron ores, and is actually Chairman of the BSI and CEN committees; he was Chairman of the now superseded ISO and PRE committees dealing with the analysis of refractory materials. Needless to say, most of this work is concerned with the XRF technique.

In the past, where chemical procedures formed the mainstay of the methods, it was not unusual for committees from different industries within the same country to produce standard methods widely different in approach for essentially the same material. Even when this was not so, there was at best considerable duplication of effort. Fortunately, with instrumental methods there is much more co-ordination of activity. Current instrumental techniques such as XRF have generally sprung from the same basic research by a few leading laboratories, so that their exploitation has followed much the same pattern in most industries, differences being usually in matters of detail. Even so, the case for a unified approach to calibration techniques over the whole pattern of materials is very strong. Methods are being proposed where the compositional ranges are very limited, so that extension of the calibration either in terms of relative contents or of additional elements to be included at a later date could involve totally new calibrations. It is possible to imagine the frustration generated to the analyst of, say, bauxites, to find totally different approaches in the standard methods for, on the one hand, aluminium-making bauxites and, on the other, refractory bauxites.

The standards organizations, fortunately, are aware of this danger, and are taking steps to overcome any such problems. For example, BSI have set up a committee to unify methods of instrumental analysis across industrial boundaries.

CALIBRATION

Also, at ISO/CEN level, a single document is being produced covering the analysis of all oxide refractories and industrial ceramics and their raw materials. This will allow alternatives in, say, fluxes and calibration techniques provided that they can be shown to produce acceptable accuracies and precisions. Such an approach is of major importance, as an analyst would not wish to maintain several sets of calibrations for the same materials or a multiplicity of short range calibrations. Neither would the analyst want to abandon a perfectly satisfactory set of calibrations and other parameters simply to conform to a national or international standard, possibly of inferior performance.

These are points which the body of XRF analysts needs to keep in mind as they follow the developing pattern of standard methods, and they should register their disapproval if the long-sighted view does not appear to be being taken.

10 Presentation of the Sample Bead and Completion of the Analysis

10.1 PRE-PRESENTATION TREATMENT

Previous chapters have taken the sample through the sample preparation process, loss on ignition determination, fusion, casting of a bead and calibrating the spectrometer. A typical analytical laboratory using XRF as its main tool will, by definition, handle large numbers of samples. This large scale activity poses problems of ensuring correct identificaton of samples, particularly at the point at which they are transferred from the preparation laboratory to the XRF laboratory. It is important, therefore, to devise a system whereby the risk of confusion is minimized.

In the authors' laboratory, the fused, cast bead would normally be transferred from the preparation laboratory to the spectrometer in a polythene re-sealable bag on which the identification number of the sample has been written. In addition, both the casting dish and the fusion dish are identified on the bag. The identity of the fusion dish is desirable so that it is possible to use the same fusion dish and include any residual melt with the bead if it has to be re-melted. This may be needed if, on visual inspection, a re-cast is thought necessary, e.g. due to bubbles or rippling. If the top surface is to be presented to the X-ray beam the identification mark on the bottom surface is produced automatically by scribing the base of the casting mould with a number or letter. This will then be moulded into the base of the bead itself, giving permanence to the labelling. Not only does this assist sample identification and good laboratory practice, but it also helps to identify any particular mould yielding problem results, say, consistently high or low totals. As well as the data on the bag, the bead should also have an adherent label on which the sample identification number is written.

Even though it has been protected from the atmosphere, and although the analyst preparing the bead will have taken care not to have touched the presentation side of the bead, it has been found that elements such as sodium, sulphur and chlorine can find their way on to the analysis surface. It is therefore desirable to rinse the presentation surface with distilled water and then to wipe it dry with a clean tissue. The bead is then immediately transferred to the holder, care again being taken to ensure that the presentation surface is not touched and that the bead lies in the correct position in the cup.

As has already been discussed, few instruments have the great advantage of the Telsec TXRF, namely that the bead, when irradiated by the X-ray beam, is held in a fixed position in the instrument itself rather than by the cup. Where, as is usually the case, the individual cup itself determines the position of the bead in relation to the X-ray beam, it is essential to ensure that none of the available cups produces any sort of bias because of slight deviation in engineering dimensions. Some instruments have more than one position inside the spectrometer into which the cups become seated. These positions similarly may differ slightly and must be checked for consistency; alternatively, only one position should be used for analysis.

10.2 PRESENTATION OF THE BEAD

The bead, having been inserted into the cup, is introduced into the instrument, either by hand or using an automatic sample changer. It is then subjected to the X-radiation using such crystals, collimators, pulse-height windows, detectors etc. and for such times as may be required for the analysis to be carried out. In addition, it is necessary to irradiate the appropriate ratio standard beads so as to provide current levels of counts for known percentages of each oxide to be analysed, which can be used to calculate ratios of sample counts to ratio standard counts. This is to avoid problems with 'drift' of the output of the instrument. It is advisable to monitor the actual number of counts obtained for each of the oxides in the ratio standard beads to confirm that the level of drift is not so great as to require attention and possible corrective measures. Also, in order to achieve the greatest certainty about the accuracy of the results, it is desirable to run a bead of a CRM or synthetic standard through the instrument in the same series as the unknown, calculating the results as though it were a sample. The results afford additional confirmation of the validity of the analysis provided that they are within experimental error of the certified results. If, however, there are discrepancies, the whole process is repeated, when hopefully such discrepancies can be shown to be merely 'sport' results. If discrepancies remain on repetition, then it is necessary to investigate and identify the problem. *It is totally incorrect to assume that the discrepancies are simply a quirk of the particular run and to adjust the sample results to fit those from the standard, however great the temptation. Regrettably this has been known to be done.*

After completion of the irradiation, the sample, CRM and ratio standard beads are returned to their respective bags for retention and for re-use as, or if, required.

PRESENTATION OF THE SAMPLE BEAD

From the above, it will be seen that to run a single sample through the instrument may require five or six beads to be irradiated, possibly each with a residence time of over 100 s in the instrument. Thus, allowing for change-over times, the instrument could be occupied in performing the single analysis for a period of about 10 minutes. This demonstrates that, whenever possible, it is highly desirable to batch similar samples together, so that the time used for running the standards etc. can be shared. An ideal batch might be, say, 12 samples. For older instruments, above this figure it becomes desirable to re-run ratio standards and CRMs to ensure consistency of the performance of the instrument. Modern instruments, however, can often maintain their stability for up to a day or even longer. The optimum size of batch should be derived in individual laboratories, based on the stability (or otherwise) of the instrument.

10.3 CALCULATION OF THE RESULTS

Having run the sample and check standard beads against the top and bottom ratio standard beads, ratios are calculated from the raw intensities (R_A). From these and the calibration curve, first estimates of percentage contents are made and used to apply α-corrections to the ratios using the equation:

$$R_T = R_A(1 + \Sigma \alpha_{ij} C_j) \tag{10.1}$$

where R_A is the apparent ratio, R_T is the 'true' ratio, α_{ij} is the α-correction of the interfering oxide (j) on the oxide being determined (i), and C_j is the concentration of the interfering oxide.

Thus for SiO_2 the equation is

$$R_{T(Si)} = R_A(1 + \alpha_{Ti} C_{Ti} + \alpha_{Al}.C_{Al} + \cdots)$$

i.e. adding the α-corrections on silica from each of the elements present.

Using the corrected ratios (R_T), the various concentrations are then calculated; these are now corrected for the effect of line interferences. For the effect of oxide j on oxide i the equation is:

$$C_T = C_A - \Sigma K_{ij} C_j \tag{10.2}$$

where C_T is the 'true' concentration of the determinand expressed as percentage content, C_A is the apparent concentration of the determinand expressed as a percentage content, and K_{ij} is the line interference correction of j on i.

Using a computer, this process, i.e. alternating the use of the two Eqs 10.1 and 10.2, is made iterative until an effectively constant value is obtained for each of the concentrations. It should be noted that, although the concentrations used for correcting are updated at the end of each loop, the corrected data (i.e. intensity ratios or drift corrected intensities) will always be the same at the start of each loop.

The results so obtained are correct, in principle, for the sample as fused; they do not take into account any contamination (WC) due to grinding, or the loss on ignition.

Modern instruments will include a computer capable of handling these calculations (and those described later) without the intervention of the analyst, the only entry required at the time of the analysis being that of the determined loss on ignition figure together with the figures for contents of oxides or elements determined by non-XRF methods, where and if appropriate. All the remaining data concerning calibration, interferences etc. will already have been entered into the memory store during the setting up of the instrument or when calibrating for a new type of sample. Sample and standard data acquired during the current run will have been automatically transferred to the computer's memory. Even where an integral computer is not available, the current cost of microcomputers with more than adequate facilities is so small that it is inconceivable that, if an XRF instrument is an economic proposition, the laboratory could not afford the small extra outlay. It is, of course, possible to compute these results using a pocket calculator, but this is a very tedious and time consuming process and is, in any event, to be strongly deprecated.

10.3.1 Correction of Errors Introduced by Tungsten Carbide Contamination

Tungsten carbide can be introduced into the sample during grinding, and is converted to tungstic oxide (WO_3) during the determination of loss on ignition. The presence of WC dilutes the sample as weighed out for the determination of loss on ignition, so that less actual sample is taken than should have been. The conversion of the carbide to the oxide is accompanied by an increase in weight, so that the measured loss on ignition is lower than it should be. Finally, the portion of sample after ignition weighed out for the fusion process is contaminated with this equivalent amount of tungstic oxide rather than carbide.

Corrections thus need to be applied as follows:

(1) to the apparent loss on ignition for (a) dilution of the sample and (b) oxidation of the carbide to the oxide;

(2) to the determinations of oxides for the dilution factor by WO_3.

The level of contamination is usually in the lower point percent range, even for hard samples, except for the very infrequent occasion when a 'chip' of tungsten carbide has become incorporated owing to damage to the mortar. Where cobalt or nickel is incorporated in the 'alloy' this is often at about the 10% level, meaning that these latter contents are unlikely to reach the 0.1% level. Thus, absolute accuracy in the level of contamination is normally of secondary importance.

PRESENTATION OF THE SAMPLE BEAD

10.3.1.1 Correction for 'Pure' Tungsten Carbide Mortars

As the degree of contamination of the sample by tungsten carbide is small, the correction of errors introduced by this factor need only to be a first approximation.

Although only two factors are being considered, viz, dilution of the sample with mortar material and the effect of this on the recorded loss on ignition, the mathematics are quite complex. The authors wish to thank their colleague, Mr F. Moore, for his derivation of the original equations.

10.3.1.2 Factor for Correcting Oxide Determinations for Loss on Ignition and WC

If the true loss on ignition expressed as a percentage is L, and the WO_3 (%) figure determined is W, then the correction for dilution is

$$\frac{100}{100-W}$$

and the correction from ignited to pre-ignited basis for the sample is

$$\frac{100-L}{100}$$

If these are combined a correction factor (F) is achieved:

$$F = \frac{100-L}{100-W} \tag{10.3}$$

It is necessary, first, to calculate the true loss on ignition from the experimental data to allow the factor F to be quantified.

10.3.1.3 Calculation of the True Loss on Ignition

The content of WO_3 (%) is determined on the ignited sample, i.e. after loss on ignition. Assuming that there is 1 g of ignited sample, this will contain $W/100$ g of WO_3 and therefore $(1 - W/100$ g$)$ of sample.

Before loss on ignition it would contain $(WR/100)$g of tungsten carbide, where R is the molecular ratio $WC/WO_3 = 0.8448$.

The weight of sample before the loss on ignition would be

$$(1 - W/100)[100/(100-L)]$$

where L is the correct loss on ignition (%).

The measured loss on ignition (M) is the loss in weight during ignition divided by the original weight of material taken expressed as a percentage, i.e.

$$M = \frac{100\{WR/100 + (1 - W/100)[100/100(100 - L)] - 1\}}{WR/100 + (1 - W/100)[100/(100 - L)]}$$

$$= 100 - \frac{100}{WR/100 + (1 - W/100)[100/(100 - L)]}$$

Rearranging gives:

$$\frac{100 - M}{100} = \frac{1}{WR/100 + (1 - W/100)[100/(100 - L)]}$$

Cross-multiplying and rearranging:

$$(1 - W/100)[100/(100 - L)] = 100/(100 - M) - WR/100$$

Thus:

$$\frac{L}{100} = 1 - \frac{1 - W/100}{\frac{100}{100 - M} - \frac{WR}{100}}$$

and:

$$L = \frac{100[100/(100 - M) - WR/100 - 1 + W/100]}{100/(100 - M) - WR/100}$$

Finally:

$$L = \frac{100[M/(100 - M) - (W/100)(R - 1)]}{100/(100 - M) - WR/100} \tag{10.4}$$

This gives the value of the true loss on ignition; this, in turn, can be substituted into 10.3 to calculate the correction factor for equations 10.5 and 10.6 below.

10.3.1.4 Factors for Cobalt or Nickel Bonded Mortars

In practice, most tungsten carbide mortars are bonded with cobalt in order to reduce the brittleness of pure tungsten carbide, but this introduces some loss of hardness. It is possible to use nickel as an alternative bonding agent if the presence of cobalt in the mortar materials needs to be avoided.

Cobalt bonding is assumed in the following discussion, but exactly the same points apply to nickel; only the conversion factors will differ. The composition of the mortar in terms of the proportions of WC and cobalt, can be derived via the determinations of WO_3 and CoO in the samples being analysed. As normal levels of contamination are generally low, the establishment of this ratio from a single sample is not usually very reliable. The determination of the ratio can be progressively refined by determining it in a succession of samples and taking an average figure. Alternatively, it occasionally happens when the mortar is wearing that it becomes chipped, thereby introducing an abnormally high level of contamination. This can be high enough to enable reliable determinations of both constituents to be made and an accurate ratio to be obtained. The best solution of all is to obtain the composition directly from the manufacturers, but not all are prepared to be co-operative. In this case, if the 'metal' of the mortar contains $Y\%$ of Co, this will give an E value of $1.0471/(100-Y)$ or, in the case of nickel, a value of $1.0750/(100-Y)$.

It will be appreciated from the above that it is very desirable to utilize mortars of the same composition throughout. If only one mortar is in use at a time, only when it is replaced will further checks on composition be required. It will, however, be clear that if several mortars are in use it is almost essential to ensure that they are all of sufficiently similar composition to demand only one set of data to be stored. The complexity of the situation where several compositions of mortar are in use is self-evident.

The ratio to be used in the calculations is that of CoO/WO_3 and not Co/WC; this is the most simple, as this is how the data arises during an analysis. This ratio may be defined as E, and it is necessary to allow for this additional dilution caused by the presence of binder in the previously derived equations. Thus, referring back to Eqn 10.3, then

$$F = \frac{100 - L}{100[1 - W/(100(1 + E))]} \quad (10.5)$$

Similarly, for calculating the true loss on ignition, Eqn 10.4 becomes:

$$L = \frac{100\{M/(100 - M) - (W/100)[(R - 1) - (EW/100)(0.7865 - 1)]\}}{100/(100 - M) - (W/100)(R + 0.7865E)} \quad (10.6)$$

The factor 0.7865 refers to the conversion of CoO to Co, the figure for nickel, Ni/NiO, is 0.7858. Bearing in mind the normally low level of contamination, a figure of 0.786 for either is adequate.

By comparison with Eqn 10.4, the $(EW/100)(0.7865 - 1)$ term compensates for the gain in weight of Co in the same way as the $(W/100)(R - 1)$ compensates for the gain in weight of the tungsten carbide. The additional $0.7865E$ term in the divisor allows for the original amount of Co in the sample just as the R value does for the tungsten carbide.

The modern DPS software of ARL and X40 software of Philips allow the above corrections to be fed into the analytical procedures, provided that the Eqns 10.5 and 10.6 are broken down. This modification and data entry will only take the analyst an hour or so.

11 Routine Techniques for Material Types

11.1 INTRODUCTION

Full experimental details and an added description covering 'know-how' are given in Chapter 8, to which general reference should be made. The present chapter is intended to indicate the layout for the chapters dealing with individual material types. As will be described, techniques which are those normally used will be described and indicated in successive chapters as 'normal'. Caveats and deviations from this norm will be described as they prove necessary for each material type.

This treatment cannot hope to be comprehensive; circumstances will arise when the analyst will be called upon to tackle either a unique type of material or a peculiar combination of elemental requirements, the exact needs of which are not covered. Even so, the general approach should give a sound lead as to how the problem should be tackled.

It will also be realized that the experimental details which follow are intended to facilitate ready reference to the needs of the material being analysed and are not complete in themselves. The data need to be used in conjunction with the more discursive information given in the earlier chapters. Also, no technology stands still, ceramics being no exception. Thus, new oxide or mineral additions may be made to specific types of material to enhance the finished product's properties; this could well demand the determination of constituents not dealt with under the appropriate heading in the later sections.

The chapters that follow give experimental details for each type of material. A full repetition of the correct conditions, in detail, for each class of material would extend the length of the book unjustifiably. For this reason these sections on specific data are preceded here by a basic description, in similar format, of the necessary data for an aluminosilicate sample, which for this purpose will be nominated as a 'normal' sample. Thus, wherever these 'normal', 'routine' or 'standard' type sample procedures apply exactly, the required conditions will merely be described as 'normal', meaning that the same treatment as for an aluminosilicate type sample is suitable. If only minor deviations from the norm are needed these are described in situ, but more significant deviations are given fuller coverage. Much of the data given for a 'normal' sample is very similar to that in Chapter 8, but the latter was meant to be more introductory and discursive.

The book covers a wide range of types of materials, but it will not, for obvious reasons, be possible to include every possible variant. Some types of almost 'one-off' material will be found, but true 'one-offs' must, by definition, require individual treatment. Analysts attempting such types will need to ensure that they have adequate experience of less demanding types and a basic understanding of the whole art (or science).

Some classes contain a range of individual materials: the most common are listed, but again this is not comprehensive. Some sections contain entries covering a whole range of materials under the generic class heading, others may contain very few types in the range, while some may be concerned with only one specific material.

The information is collected under appropriate process headings, and an indication of the sort of data contained in each is given. This is followed in each section by the conditions which will subsequently be described as 'normal' and to which reference may need to be made.

11.2 'NORMAL' EXPERIMENTAL CONDITIONS

11.2.1 Typical Materials

Where there are several types of material in the general classification those most commonly encountered are listed.

11.2.2 Sample Preparation

After appropriate particle size reduction and sample bulk reduction, final grinding is carried out in tungsten carbide (Chapter 2). The particle size of the material should be reduced so that most passes through a 125 μm mesh. At this fineness most materials will fuse readily in the appropriate flux.

This topic will be specifically dealt with only if there is some specific problem. The general problems arising from grinding hard, abrasive materials (particularly where small contents are to be measured) and soft 'sticky' materials have been dealt with in Chapter 2. Each sample has to be dealt with on its merits, taking into account the equipment available. Double sample preparations by different routes may sometimes prove to be the only solution.

11.2.3 Drying

The sample is dried in an air oven at $110 \pm 5°C$ for a minimum of 2 hours, but a more common time is 4 hours; overnight is usually permissible. However, prolonged drying of samples containing combined water is potentially undesirable.

Deviations will be indicated and reasons given.

11.2.4 Weighing

All weighing should be carried out with the minimum of delay, as almost all finely ground powders, irrespective of composition, will tend to absorb water on the exposed surface. Weigh rapidly to within 0.5 mg of the desired weight.

Some samples tend to be hygroscopic or to pick up carbon dioxide from the atmosphere either before or after ignition. Those requiring care in the former case will be noted under this heading. The treatment of the latter will be mentioned under the heading of loss on ignition.

11.2.5 Loss on Ignition

Weigh $1.000 \text{ g} \pm 0.5$ mg of sample into a suitable platinum or platinum alloy crucible or $2.000 \text{ g} \pm 0.5$ mg into the platinum/gold (or other suitable) dish to be used for the fusion. The latter alternative is preferable, and should be used except when sintering of the sample may be expected to be so severe that adequate mixing of the sample and flux, at a later stage, would not be possible. In the latter situation it is often best to use a porcelain crucible, which is discarded after use.

Almost cover the crucible or dish with a lid and follow one of the following procedures.

(i) Place the crucible or dish into a furnace and slowly raise the temperature to $1025 \pm 25°C$, maintaining an oxidizing atmosphere throughout. The minimum time for this increase in temperature is 20 min and the ultimate temperature should be maintained for 30 min.

(ii) Start the ignition over a low flame, slowly increasing to a dull red heat over a period of about 20 min, then transfer the crucible/dish and contents into a furnace at $1025 \pm 25°C$, maintaining an oxidizing atmosphere throughout. Maintain at this temperature for 30 min.

(iii) Place the crucible/dish at the entrance to a suitably programmed tunnel kiln to carry out the increase in temperature and subsequent ignition at $1025 \pm 25°C$.

Remove the crucible/dish from the furnace, completely cover with the lid, cool to room temperature in a desiccator and then re-weigh immediately. Silica gel is usually adequate as a desiccant.

Ignition for 30 min is usually sufficient, but confirmation of this should be made with unfamiliar types of sample. This entails weighing the cooled, ignited sample, re-igniting for a further 30 min and re-weighing.

Deviations from the norm will be indicated as necessary, but problems requiring discussion can arise either from the type of material or, possibly, from the presence of a specific oxide.

11.2.6 Fusion

Flux: 4 parts by weight of lithium metaborate:1 part by weight of lithium tetraborate.

Flux/sample ratio: 5/1; 7.500 g ± 1 mg of flux/1.500 g ± 0.5 mg of sample. These figures are those to produce a 35 mm diameter bead.

Fusion Procedure

The procedure described below is that adopted by the authors and is within the specification of the Standard Method in BS 1902.

> *(i) Using the ignited sample from the loss on ignition*
> Remove sufficient ignited sample from the dish to leave 1.500 g ± 0.5 mg in the dish for the fusion.
>
> *(ii) Using un-ignited sample*
> Calculate the weight of unignited sample (W g) required to yield the equivalent of 1.500 g of ignited sample. Weigh W g ± 0.5 mg of the sample into the dish. (More care must be taken during the earlier stages of the fusion, with this option, to ensure that all carbonaceous matter is burnt out and that all iron compounds are converted to the ferric state. This must be complete before proceeding above dull red heat.)

Carefully mix the sample and flux uniformly with a platinum wire or, more usually, with a stainless steel or nickel spatula. (Take care, if trace elements are to be sought.) Almost cover the dish with a lid and heat over a low gas burner. Gradually increase the temperature to the full heat of the burner over a period of about 10 min and continue heating. Unless an automatic swirling device is available, swirl frequently to ensure thorough mixing of the melt once the flux has completely melted. Ensure, by visual inspection, that decomposition appears to be complete, then transfer the dish and contents to a furnace at 1200°C, covering the dish completely with a lid. Also, at the same time, introduce into the furnace a casting mould standing on a heat reservoir (a 50 mm × 50 mm piece of sillimanite, 10 mm thick, is suitable). The furnace should have a positively oxidizing atmosphere throughout. If the top surface of the bead is to be used, the casting mould may be marked internally with a number or letter which will be transferred to the bead to assist in sample identification at later stages of the analysis. This marking may be made by indenting a number or letter into the base by means of a stamp or, better, by engraving, e.g. with a diamond drill.

It will be seen that this section is split into the three headings flux type, flux/sample ratio and fusion procedure. The first two are given in full, but deviations or different procedures are spelled out.

11.2.7 Casting

After a 5 minute period, remove the heat reservoir and casting mould from the furnace and place them on a flat, horizontal surface; immediately remove the lid from the fusion dish, rapidly swirl to assist final mixing and pour the melt from the dish into the casting mould. As much of the melt as possible should be transferred to the casting mould to ensure minimum deviation in the curvature at the top of the bead. This is best achieved by just touching the last drop of the melt to the edge of the melt already in the dish. Thus, any slight irregularity will probably lie outside the area irradiated by the X-ray beam.

The melt should then be transferred to a position above an air jet placed centrally to the melt. This can be done either before the melt has started to solidify, in which case the casting mould should be arranged horizontally (giving a bead of uniform cross-section through any diameter), or immediately after solidification. The latter technique is essential if there is a danger that such movement during solidification could result in distortion of the bead. On the other hand, the former is preferable if the type of sample is liable to produce a bead that tends to devitrify, e.g. high aluminas, and samples containing significant levels of magnesia, zirconia or phosphorus pentoxide. The solidified bead should separate cleanly from the platinum alloy on cooling; if sticking becomes apparent, it is permissible to tap the mould *gently* to assist release as soon as Newton's rings appear at the base of the casting dish. After cooling to room temperature, it is advisable to stick on to the back of the bead an identifying label and place the whole in a re-sealable polythene bag on which is written identification data (i.e. sample number, fusion dish number and casting mould identification). This information may be very valuable for any possible post mortem.

The above procedure is virtually universal.

11.2.8 Oxides to be Determined

It is intended to deviate from the use of 'normal' in this section. As far as practicable, a list of usually determined and occasionally required constituents, whether determined by XRF or otherwise, will be given in this section for each type of material. This list is based on experience, but cannot be fully comprehensive, as specific circumstances can require the determination of extra constituents not normally called for.

In the information concerning specific types of material the constituents will be separated into three categories: those normally determined by XRF in the course of a routine analysis; those that may be required in some special circumstances (this cannot be exhaustive for obvious reasons); and non-XRF determinable constituents. These may or may not be required, depending on the information sought.

11.2.9 Calibration Ranges

The figures for aluminosilicates and the other types are given in each case in the following chapters on specific materials.

Where the type of material makes this sensible, normal calibration ranges are given for each element in the usually required range and, occasionally, those others that may be required less frequently. Except where specifically mentioned, calibration ranges are given as they can be expected to be met with in normal practice rather than covering infrequent unusual additions. Nevertheless, those included will cover normal additions made for ceramic reasons as well as levels found naturally.

A distinction needs to be drawn between three possible lower limits to be considered as the bottom of the calibration range.

First, there is the *calibration range* for each oxide itself, that is the range of contents, expressed in percentage terms in the sample, of the oxide added to produce the standard bead. For the bottom point in this case the amount added is obviously zero, giving a figure of 0%.

Second, there is the *reporting limit*, that is the lowest amount present of the oxide reported, and below which the less-than sign (<) is shown. This is commonly 0.01%, but may be higher for say, Na_2O, where a higher figure has to be used because of lack of sensitivity, or lower where the lower figure can be achieved and is advantageous.

Lastly, there is the *'limit of detection'*, this being the smallest amount of the oxide that can be positively identified. This figure is interpreted differently by different schools of thought. Some regard 2σ (σ being the standard deviation of repeated results) as being an appropriate figure, whereas others insist on 3σ. The *reported* result should, in the authors' opinion, be put at a figure that can be regarded as certain, that is, sufficiently above the more stringent detection limit to ensure that the user of the analysis is not deceived.

11.2.10 Reporting of Results

Results are normally reported as the oxides in the oxidation states shown by the formulae in Chapter 7.

Results are normally reported on the dried (110°C) basis, but with certain types of sample, or for reasons associated with the specific data required, the results may be reported on the ignited (1025°C) basis, or even 'as received'. In the former case it is normal to report the determined result for the loss on ignition separately.

Sulphur trioxide:	Before ignition, entered separately from the analysis; after ignition, added in as SO_3.
Halogens:	Reported as element (e.g. F, Br, Cl); the equivalent percentage of oxygen needs to be deducted from the analytical total as 'oxygen equivalent of fluorine' etc.
Carbon:	The figure entered is usually that obtained by ignition (pyrolysis), from which the carbon content from the separately determined carbon dioxide figure has been deducted. The carbon result is reported separately.
Carbon dioxide:	Normally reported separately.
Loss on ignition:	There may, on occasion, be an increase in weight due to oxidation, say, of ferrous iron to ferric. This should be reported as 'gain on ignition' and the figure *deducted* from the analytical total.

In some cases it is not appropriate to report the results as oxides, and in others a specific oxide may need to be reported in an unusual oxidation state. Deviation from routine will be given where necessary.

Iron may well be present in a variety of oxidation states, tramp iron, ferrous oxide and ferric oxide. Fe_3O_4 may also be present, but this ferroso-ferric oxide may be expressed as $FeO \cdot Fe_2O_3$. Unless specific determinations are made of the other oxidation states the whole of the iron present is normally reported as ferric oxide. This should correctly be expressed as 'Total iron as ferric oxide' on the analytical report, but this is often taken as read when the figure for Fe_2O_3 is quoted. If other oxidation states are separately reported these figures are usually placed underneath the analytical total or the various contents included in the analytical list, bearing in mind that, in the latter case, the loss on ignition will need to be corrected for the oxygen used to convert to the ferric state.

12 Procedures for Silica/Alumina Range Materials

This range can be accommodated on one set of calibrations if so desired, but it is better to calibrate separately lower levels of silica and alumina, where appropriate, to ensure adequate levels of accuracy. However, some laboratories will be concerned with more limited applications, and will therefore be better advised to calibrate over the separate ranges actually used, or to use that portion of the whole calibration within which their types of samples fall. Thus, although this section takes account of the full range, the various portions of the range, where they call for modified treatments, are also dealt with as may be required under each heading.

The sub-classifications are, then: (A) Aluminosilicate Materials, (B) Aluminous Materials, (C) High-Silica Materials, and (D) High-Alumina Materials.

12.1 TYPICAL MATERIALS

(A) Aluminosilicate Materials

Clay	ball clay
	china clay (kaolinite)
	fireclay
	illitic
	marl
	bentonite (montmorillonite)
	shale
Feldspar	albite (soda feldspar)
	orthoclase (potash feldspar)
	cornish stone (needs P_2O_5 and F determinations)
	petalite (high lithia)
Nepheline Syenite	
Mica	muscovite
	illite
	lepidolite (high lithia)

Manufactured Products	pottery: earthenware, stoneware, porcelain.
	refractories: firebrick, molochite.
Heavy Clay (Building Products)	bricks, tiles,
	clay pipes.

(B) Aluminous Materials

As far as ceramics are concerned, these materials are used almost exclusively by the refractories industries. They are: kyanite, andalusite, sillimanite, flint clay, mullite and bauxite (lower purity types).

(C) High-silica Materials

For many high-silica materials, it is possible to ignore interference corrections if specific calibrations for the class are made, except where ultimate accuracies are required. Most analysts, however, who may be using broad-based calibrations and have corrections built into the computer system, will prefer to utilize them.

If any of the determinations fall into the trace element range, special XRF methods or other techniques may be needed.

Pottery	flint
	sand (glass industry also)
	cristobalite
Refractories	ganister
	quartz
	quartzite
	silica brick
	fused silica (CARE: possible specks of silicon)
Heavy Clay	sand

(D) High-alumina Materials

Such materials are: bauxite (refractory), mullite and prepared aluminas.

Note: High-alumina materials, when ground, may adsorb water from the atmosphere. It is not unusual for an alumina, possibly previously heated to 1600°C or even fused, to show a loss on ignition of 0.5–1% on heating to 1025°C. Thus, it may well be preferable to calculate the analysis on the ignited basis rather than on the 110°C dried basis.

12.2 SAMPLING

(A) Aluminosilicate Materials

Normal.

(B) Aluminous Materials

Normal. These higher alumina content materials are generally very tough, causing greater levels of contamination during grinding than most materials.

(C) High-silica Materials

Nomal. Some very pure types need great care and changed procedures to avoid significant contamination.

(D) High-alumina Materials

Normal. These materials, if in massive form, are very difficult to grind, and may produce much higher than normal levels of contamination.

12.3 DRYING

(A) Aluminosilicate Materials

Normal. Some clays, particularly bentonites (montmorillonite), lose water progressively at any ordinary drying temperature. Drying of these types needs to be restricted both to minium and maximum times. Specified drying times should be adopted consistent with the purposes for which the data is required.

(B) Aluminous Materials; (C) High-silica Materials

Normal.

(D) High-alumina Materials

Normal. High-alumina materials can also retain water to high temperatures, so that not all will be removed by drying at 110°C.

12.4 WEIGHING

(A) Aluminosilicate Materials

Normal. Some clays are hygroscopic and need to be weighed rapidly.

(B) Aluminous Materials; (C) High-silica Materials; (D) High-alumina Materials

Normal.

12.5 LOSS ON IGNITION

(A) Aluminosilicate materials

Normal. For unignited samples, burn off carbon and oxidize iron to the ferric state by slowly raising the temperature.

(B) Aluminous Materials; (C) High-silica Materials

Normal.

(D) High-alumina Materials

Normal. With some high-alumina materials, as was noted under the section on drying, some adsorbed water will be driven off during the ignition. This water has been adsorbed only as a result of grinding, i.e. from the increase in surface area. This water was not present in the sample before grinding, so that the analysis of these latter materials is best recorded on the ignited 1025°C basis, with the loss on ignition being recorded separately, for information.

12.6 FUSION

12.6.1 Flux

All Types (A, B, C and D)

4 lithium metaborate:1 lithium tetraborate.

12.6.2 Flux/Sample Ratio

5:1. $7.500 \text{ g} \pm 1 \text{ mg}$ of flux : $1.500 \text{ g} \pm 0.5 \text{ mg}$ of ignited sample for a 35 mm diameter bead.

(C) High-silica Materials

Normal.

Note: Better sensitivities may be obtained with 'pure' samples by weighing several grams of the material into a dish and removing the silica by hydrofluoric/sulphuric acid treatment and ignition. If a further $1.500 \text{ g} \pm 0.5 \text{ mg}$ is weighed into the dish and the fusion etc. is continued, the impurities from the total weight can be calculated as percentages. Using this technique, for favourable oxides, ppm levels are determinable.

12.6.3 Fusion Procedure

(A) Aluminosilicate Materials

Ignited samples, as for standard method; with unignited samples care needs to be taken to burn off carbon and oxidize iron to the ferric state at lower temperatures.

(B) Aluminous Materials

Normal, but essential to use 1200°C furnace to complete.

(C) High-silica Materials

Normal.

(D) High-alumina Materials

Normal. In this case, the final fusion at 1200°C is imperative to ensure the final removal of minute traces of unattacked corundum. If this is not decomposed it can act as a nucleating agent, causing a tendency for the bead to devitrify.

12.7 CASTING

All Types (A, B, C and D)

Normal.

12.8 CONSTITUENTS TO BE DETERMINED

12.8.1 By XRF

All Types (A, B, C and D)

Silica, alumina, ferric oxide, titania, lime, magnesia, soda, potash. Phosphorus pentoxide and zirconia may also be required.

(A) Aluminosilicates

Clays: manganic oxide and barium oxide are often determined. Vanadium pentoxide may also be relevant in specific cases.

Pottery Bodies: cobalt (added to decolorize the ware) may occasionally be needed.
Feldspars: rubidia, caesia and strontia may be called for.

(B) Aluminous Materials

Chromium sesquioxide, manganic oxide, gallium oxide, zinc oxide and vanadium pentoxide may occasionally be requested.

(C) High-silica Materials

Chromium sesquioxide may be required for glass sands, but will normally be at the lower ppm level, and thus needs determining by other techniques, e.g. colorimetrically.

(D) High-alumina Materials

Chromium sesquioxide and gallium oxide (often found at about the 0.015% level) may be determined.

12.8.2 Constituents Determined by Other Methods

All Types (A, B, C and D)

Lithia is present in many of these materials at lower than 0.1% level, and may on occasion be called for. It has, of course, to be determined in lithium ores, and may be deliberately added to some glasses and glazes that can be analysed using these calibrations.

Boric oxide, usually at low levels, on materials required for use in the atomic energy industry.

Occasionally in ceramic work (glasses) and normally in geochemical analysis, the determination of ferrous oxide may be required.

(A) Aluminosilicate Materials

Clays
Sulphur trioxide before and/or after ignition. The latter figure may often be obtained with sufficient accuracy from the XRF analysis.
Carbon: as an assessment of carbonaceous matter.
Carbon dioxide, from carbonates.
Fluorine.
Chlorine may be needed on 'clays' from the Middle East.

12.9 CALIBRATION RANGES

Complete Alumina–silica Range

		%
Silica	SiO_2	0–100
Titania	TiO_2	0–5
Alumina	Al_2O_3	0–100
Ferric Oxide	Fe_2O_3	0–12
Lime	CaO	0–16
Magnesia	MgO	0–10
Soda	Na_2O	0–12
Potash	K_2O	0–20

(A) Aluminosilicate Materials

		%
Silica	SiO_2	40–90
Titania	TiO_2	0–5
Alumina	Al_2O_3	5–50
Ferric Oxide	Fe_2O_3	0–12
Lime	CaO	0–16
Magnesia	MgO	0–10
Soda	Na_2O	0–12
Potash	K_2O	0–20
Phosphorus pentoxide	P_2O_5	0–5

(B) Aluminous Materials

		%
Silica	SiO_2	0–50
Titania	TiO_2	0–8
Alumina	Al_2O_3	20–100
Ferric oxide	Fe_2O_3	0–40*
Lime	CaO	0–5
Magnesia	MgO	0–5
Soda	Na_2O	0–5
Potash	K_2O	0–5
Phosphorus pentoxide	P_2O_5	0–5

*0–12% will probably be adequate if non-refractory bauxites are not be analysed.

(C) High-silica Materials

		%
Silica	SiO_2	90–100
Titania	TiO_2	0–5
Alumina	Al_2O_3	0–5
Ferric oxide	Fe_2O_3	0–5
Lime	CaO	0–5
Magnesia	MgO	0–5
Soda	Na_2O	0–5
Potash	K_2O	0–5

(D) High-alumina Materials

		%
Silica	SiO_2	0–10
Titania	TiO_2	0–5
Alumina	Al_2O_3	85–100
Ferric oxide	Fe_2O_3	0–5
Lime	CaO	0–5
Magnesia	MgO	0–5
Soda	Na_2O	0–5*
Potash	K_2O	0–5

*Some aluminas contain low levels of soda, Na_2O, the exact amount being of great technological significance. Unless a modern spectrometer is available, capable of the required sensitivity, other methods of analysis may need to be used.

12.10 REPORTING OF RESULTS

All Types (A, B, C and D)

Normal.

13 Procedures for Calcium-rich Materials

Calcium may be found in combination with major amounts of several oxides or elements. All are handled in this chapter: 1, with silica, 2, with silica and phosphate, 3, with phospate, 4, with sulphate, 5, with fluoride, and 6, with carbonate with/without magnesium.

13.1 CALCIUM SILICATES

13.1.1 Typical Materials

Portland cement, anorthite, calcium silicates.

13.1.2 Sampling

Normal. Tends to stick to surfaces of the grinding vial. Make sure the sample is dry. The vial, puck and ring may need to be cleaned by 'wet' grinding with a clean sand.

13.1.3 Drying

Normal. Over-long drying should be avoided; with some materials in the range there is a serious risk of reaction with moisture and carbon dioxide from the atmosphere.

13.1.4 Weighing

Normal.

13.1.5 Loss on Ignition

Normal, but note that some national standard methods specify temperatures other than 1025°C. With normal Portland cements virtually all the sulphur trioxide will remain after ignition.

13.1.6 Fusion

13.1.6.1 Flux

4 lithium metaborate:1 lithium tetraborate.

13.1.6.2 Flux/Sample Ratio

5:1. 7.500 g ± 1 mg of flux:1.500 g ± 0.5 mg of ignited sample for a 35 mm diameter bead.

Note: For those analysing only cements and limestones, the samples may be analysed after fusing in lithium tetraborate, with a 1:5 sample to flux ratio, and using extended limestone calibrations. This would help to minimize the number of calibrations.

13.1.6.3 Fusion Procedure

Normal. With care, and using 4:1 flux and a lower final temperature, sulphur trioxide can be retained during fusion, and reasonable results for this constituent can be obtained on the bead by XRF. On the contrary, sulphide sulphur, if present, is often lost.

13.1.7 Casting

Normal.

13.1.8 Constituents to be Determined

13.1.8.1 By XRF

Silica, alumina, ferric oxide, titania, lime, magnesia, soda and potash. Phosphorus pentoxide, barium oxide, strontia, manganic oxide and chromium sesquioxide may also be required. Sulphur trioxide may often be obtained by XRF.

13.1.8.2 By Other Methods

Sulphur trioxide (but see comments above), chloride (the presence of excessive chloride in cements can damage metal reinforcement in concrete), and carbon dioxide.

13.1.9 Calibration Ranges

		%
Silica	SiO_2	0–50
Titania	TiO_2	0–5

PROCEDURES FOR CALCIUM-RICH MATERIALS

		%
Alumina	Al_2O_3	0–20
Ferric oxide	Fe_2O_3	0–20
Lime	CaO	30–70
Magnesia	MgO	0–10
Soda	Na_2O	0–5
Potash	K_2O	0–5
Sulphur trioxide	SO_3	0–5

13.1.10 Reporting of Results

Normal, but sulphur trioxide after ignition (in the bead) should be added into the analytical total.

13.2 CALCIUM WITH SILICA AND PHOSPHATE (BONE CHINA BODIES)

13.2.1 Typical Materials

Pottery bodies of normal bone china composition (traditionally 50% bone ash, 25% clay, 25% flux), although there are some variations.

13.2.2 Sampling

Normal.

13.2.3 Drying

Normal.

13.2.4 Weighing

Normal.

13.2.5 Loss on Ignition

Normal.

13.2.6 Fusion

13.2.6.1 Flux

4 lithium metaborate:1 lithium tetraborate.

13.2.6.2 *Flux/Sample Ratio*

5:1. 7.500 g ± 1 mg of flux : 1.500 g ± 0.5 mg of ignited sample for a 35 mm diameter bead.

13.2.6.3 *Fusion Procedure*

Normal. High phosphate content demands care to prevent grain formation in the platinum alloy of the dish. Maintain good oxidizing conditions throughout.

13.2.7 Casting

Normal.

13.2.8 Constituents to be Determined

13.2.8.1 *By XRF*

Silica, alumina, ferric oxide, titania, lime, magnesia, soda, potash and phosphorus pentoxide. Barium oxide, strontia, manganic oxide and chromium sesquioxide may also be required.

13.2.9 Calibration Ranges

		%
Silica	SiO_2	25–35
Titania	TiO_2	0–5
Alumina	Al_2O_3	0–20
Ferric oxide	Fe_2O_3	0–5
Lime	CaO	15–30
Magnesia	MgO	0–5
Soda	Na_2O	0–5
Potash	K_2O	0–5
Phosphorus pentoxide	P_2O_5	15–30

13.2.10 Reporting of Results

Normal, but including phosphorus pentoxide as standard.

13.3 CALCIUM/PHOSPHORUS PENTOXIDE

13.3.1 Types of Material

Bone ash, apatite, synthetic tricalcium phosphate, synthetic dicalcium phosphate, phosphate rock (e.g. for fertilizer).

13.3.2 Sampling

Normal.

13.3.3 Drying

Normal.

13.3.4 Weighing

Normal.

13.3.5 Loss on Ignition

Normal. As the material may contain various apatites, including hydroxy-carbonato- and sulphato-, these may not be totally decomposed at 1025°C. Depending on the purpose of the analysis, it may be necessary to increase the ignition temperature to 1200°C or to carry out determinations of residual H_2O, CO_2, SO_3 etc.

13.3.6 Fusion

13.3.6.1 Flux

4 lithium metaborate:1 lithium tetraborate.

13.3.6.2 Flux/Sample Ratio

5:1. $7.500 \, g \pm 1 \, mg$ of flux:$1.500 \, g \pm 0.5 \, mg$ of ignited sample for a 35 mm diameter bead.

13.3.6.3 Fusion Procedure

Normal.

13.3.7 Casting

Normal.

13.3.8 Constituents to be Determined

13.3.8.1 By XRF

Silica, alumina, ferric oxide, titania, lime, magnesia, soda, potash, phosphorus pentoxide, barium oxide, and strontia. Other elements may be required at trace levels; these will usually depend on the purpose of the analysis. Sulphur trioxide remaining after the loss on ignition determination may sometimes be obtained by XRF measurement, but the acidity of the phosphorus pentoxide is less favourable to the retention of sulphur trioxide.

13.3.8.2 Constituents Determined by Other Methods

Sulphur trioxide (see above); carbon, to check degree of calcination; this figure may be unreliable if the carbonate content is large in relation to the carbon content; carbon dioxide from carbonates; fluorine; boric oxide, very infrequently, usually at low levels: it has been suspected of having an influence on the liquid properties of the bone slop.

13.3.9 Calibration Ranges

		%
Silica	SiO_2	0–5
Titania	TiO_2	0–5
Alumina	Al_2O_3	0–5
Ferric oxide	Fe_2O_3	0–5
Lime	CaO	25–65
Magnesia	MgO	0–5
Soda	Na_2O	0–5
Potash	K_2O	0–5
Phosphorus pentoxide	P_2O_5	35–75

13.3.10 Reporting of results

Normal, plus phosphorus pentoxide. Depending on the ignition temperature and the needs of the analysis, water (hydroxyl), carbon dioxide, sulphur trioxide and fluorine may also need to be reported. Using the lower temperature of ignition, they may need to be considered in the total, bearing in mind that some of each may have been removed by the ignition.

13.4 CALCIUM/SULPHUR TRIOXIDE

13.4.1 Typical Materials

Gypsum ($CaSO_4 \cdot 2H_2O$), plaster (of Paris) ($CaSO_4 \cdot 0.5H_2O$), anhydrite ($CaSO_4$).

13.4.2 Sampling

Normal. Care should be taken when grinding hydrated forms of calcium sulphate

PROCEDURES FOR CALCIUM-RICH MATERIALS 207

that the temperature does not rise significantly, otherwise some water of crystallization may be lost (see Drying). Will tend to stick in the grinding vial.

13.4.3 Drying

Drying at temperatures of 110°C will result in the loss of water of crystallization. If a complete analysis is required, including an estimate of the state of crystallization, it may be advisable to 'dry' below 40°C, obtaining a figure for 'water' via the loss on ignition figure or by a specific determination. It may even be appropriate to report the results on an 'as-received' basis.

13.4.4 Weighing

Normal.

13.4.5 Loss on Ignition

The loss on ignition is carried out at 550°C, this being a compromise temperature allowing the removal of water and carbonaceous material, but not initiating any decomposition of sulphates or carbonates. This means that a determination of CO_2 needs to be carried out.

13.4.6 Fusion

13.4.6.1 Flux

4 lithium metaborate:1 lithium tetraborate.

13.4.6.2 Flux/Sample Ratio

5:1. $7.500 \text{ g} \pm 1 \text{ mg}$ of flux:$1.500 \text{ g} \pm 0.5 \text{ mg}$ of ignited sample for a 35 mm diameter bead.

13.4.6.3 Fusion Procedure

Fusion is carried out at 1050°C, and the analysis may include the determination of residual SO_3. Given adequate care, this latter figure is usually accurate enough to be used as the content of SO_3 in the sample. Bearing in mind the complexity of the chemical procedure, the results from XRF may be of equal accuracy to chemical figures.

13.4.7 Casting

Normal.

13.4.8 Constituents to be Determined

13.4.8.1 By XRF

Silica, alumina, ferric oxide, titania, lime, magnesia, soda and potash. Phosphorus pentoxide, barium oxide and strontia may also be required, plus possibly sulphur trioxide.

13.4.8.2 By Other Methods

Sulphur trioxide may need to be determined by alternative methods, such as by precipitation as barium sulphate with a gravimetric finish; carbon dioxide from carbonates; fluorine may be infrequently required.

13.4.9 Calibration Ranges

		%
Silica	SiO_2	0–5
Titania	TiO_2	0–5
Alumina	Al_2O_3	0–5
Ferric oxide	Fe_2O_3	0–5
Lime	CaO	30–50
Magnesia	MgO	0–5
Soda	Na_2O	0–5
Potash	K_2O	0–5
Sulphur trioxide	SO_3	40–60

13.4.10 Reporting of Results

Normal; the figure for sulphur trioxide is placed after the normal oxides and added into the total.

13.5 CALCIUM/FLUORINE

13.5.1 Typical Material

Fluorspar.

13.5.2 Sampling

Normal; may tend to stick in the grinding vial.

13.5.3 Drying

Normal.

PROCEDURES FOR CALCIUM-RICH MATERIALS

13.5.4 Weighing
Normal.

13.5.5 Loss of Ignition
Normal. Unless the sample contains a large siliceous impurity, when SiF_4 may be lost, there is no problem. There may be a tendency for the material to melt.

13.5.6 Fusion

13.5.6.1 Flux
4 lithium metaborate:1 lithium tetraborate.

13.5.6.2 Flux/Sample Ratio
5:1. 7.500 g ± 1 mg of flux:1.500 g ± 0.5 mg of ignited sample for a 35 mm diameter bead.

13.5.6.3 Fusion Procedure
Fusion should be carried out at 1050°C. Most of the fluorine can be retained, but the melt needs to be weighed and the appropriate correction made.

13.5.7 Casting
Normal.

13.5.8 Constituents to be Determined

13.5.8.1 By XRF
Silica, alumina, ferric oxide, titania, lime, magnesia, soda, potash, barium oxide and strontia. Phosphorus pentoxide may also be required. Lead oxide may occasionally be required.

Fluorine: Modern spectrometers can make a reasonably accurate assessment of the content of fluorine. If sufficient for the needs of the analysis, the determination of fluorine by XRF may be possible, but this point needs to be checked. It is advisable to confirm the level of accuracy achieved by cross-checking with results from chemical methods for a number of samples before relying on the results.

13.5.8.2 By Other Methods
Sulphur trioxide; carbon dioxide from carbonates; fluorine, unless this has been shown to be determinable by XRF.

Soluble CaO: This figure is often required to evaluate the amount of 'calcium carbonate' present in the sample in order to allow the 'calcium fluoride' to be calculated. This provides the technologist with a partial 'rational' analysis of the material and enables the purity or otherwise of the material to be assessed. Fluorspar is often used in the manufacture of steel, and this information is then required.

13.5.9 Calibration Ranges

		%
Silica	SiO_2	0–5
Titania	TiO_2	0–5
Alumina	Al_2O_3	0–5
Ferric oxide	Fe_2O_3	0–5
Lime	CaO	50–70
Magnesia	MgO	0–5
Soda	Na_2O	0–5
Potash	K_2O	0–5
Barium oxide	BaO	0–5
Strontia	SrO	0–5
Fluorine	F	0–50

13.5.10 Reporting of Results

Normal. Fluorine determined by other methods (or by XRF) should be added in, but with an additional line showing the oxygen equivalent of the fluorine, which is deducted from the total. This is to allow for the fact that the calcium has been entered as the oxide, thereby adding in oxygen which is not present. Alternatively, the fluorine may be calculated as CaF_2 and the equivalent CaO deducted from the determined result. This is not a verifiable assumption, as some of the barium and strontium may be present as fluorides.

13.6 CALCIUM/CARBON DIOXIDE WITH/WITHOUT MAGNESIUM: DOLOMITE AND LIMESTONE

NOTE: If the laboratory only analyses limestones, dolomites and magnesites, then economy of effort may justify combining all three types on a single calibration using a flux to sample ratio of 10:1 rather than 5:1.

13.6.1 Typical Materials

Dolomite, dolomitic limestone, limestone, quicklime (CaO), slaked lime [$Ca(OH)_2$], chalk, Iceland spar, calcium carbonate.

13.6.2 Sampling

Normal. Most of these materials tend to stick badly to the sides of the vial and the puck and ring when a Tema mill is used.

13.6.3 Drying

Normal, except when the sample contains hydrated material. Much of this water may be lost on drying to 110°C. Drying conditions will therefore depend on the data required of the analysis, and separate determination of the H_2O may be necessary. Care needs to be taken to avoid carbon dioxide being absorbed by some of the materials in this section.

13.6.4 Weighing

Normal. Materials such as quicklime will absorb water from the atmosphere during weighing, and both quicklime and hydrated materials will tend to absorb carbon dioxide from the atmosphere. Rapid weighing may be adequate in many cases but, on occasion, weighing out a dead weight may be necessary. This is best achieved by removing the portion to be analysed from a stoppered weighing bottle which is weighed before and after the removal. As with the drying procedure, the technique adopted must depend both on the nature of the sample and the information required from the analysis.

13.6.5 Loss on Ignition

Normal. The ignited material may well be vigorously hygroscopic and also capable of absorbing carbon dioxide from the atmosphere. Weighing of the ignited material should be rapid, and the crucible or dish should have a well fitting cover. In addition, it should be remembered that the material may well contain high levels of content of H_2O and/or CO_2. Thus, care should be taken to avoid over-rapid increase in temperature, otherwise spurting may occur. It also means that it is advisable to repeat the ignition (and weighing) to ensure the attainment of constant weight.

13.6.6 Fusion

13.6.6.1 Flux

Lithium tetraborate.

13.6.6.2 Flux/Sample Ratio

5:1. 7.500 g ± 1 mg of flux : 1.500 g ± 0.5 mg of ignited sample for a 35 mm diameter bead.

It is often better with these classes of materials, particularly if they are in the raw state, to carry out a separate loss on ignition and weigh out a weight of the raw sample so as to provide 1.500 g of the ignited material for subsequent fusion. This avoids the difficulty of weighing out an exact weight of ignited sample, which may be rapidly absorbing H_2O or CO_2 from the atmosphere. If this course is followed, care needs to be taken with high-loss samples to prevent spurting.

13.6.6.3 Fusion Procedure

Normal.

13.6.7 Casting

Normal. The greater viscosity of the tetraborate needs to be borne in mind.

13.6.8 Constituents to be Determined

13.6.8.1 By XRF

Silica, alumina, ferric oxide, titania, lime, magnesia, soda, potash, phosphorus pentoxide and barium oxide. Chromium sesquioxide, manganic oxide and strontia may also be required.

13.6.8.2 By Other Methods

Sulphur trioxide; carbon, as an assessment of carbonaceous matter. This is particularly significant for tar-bonded refractories. The carbon in the carbonate, calculated as C, needs to be deducted from the 'carbon' figure determined by combustion. The possibly high content of carbonate can give rise to great difficulty in determining the 'carbon' figure with any degree of accuracy. 'Tar' determinations are often carried out by means of an extraction with an organic solvent; carbon dioxide from carbonates.

13.6.9 Calibration Ranges

		%
Silica	SiO_2	0–5
Titania	TiO_2	0–5
Alumina	Al_2O_3	0–5
Ferric oxide	Fe_2O_3	0–5

		%
Lime	CaO	50–100
Magnesia	MgO	0–50
Soda	Na_2O	0–5
Potash	K_2O	0–5
Barium oxide	BaO	0–5
Strontia	SrO	0–5
Phosphorus pentoxide	P_2O_5	0–5

13.6.10 Reporting of Results

Normal. Barium oxide and strontia should be added in. For raw materials, if appropriate, the hydroxide, carbon dioxide and sulphur trioxide figures may need to be reported separately.

14 Procedures for Magnesium-rich Materials

Magnesium-rich materials include oxides, hydroxides, carbonates and silicates. These may conveniently be divided into two classes, with all but the last being analysed in a similar manner. It is also convenient to include in this chapter materials in which significant contents of chrome are included with magnesia, e.g. chrome refractories and ores.

14.1 MAGNESIUM HYDROXIDES, OXIDES OR CARBONATES

14.1.1 Typical Materials

Magnesium carbonate, magnesia, sea-water magnesia, magnesite, magnesia-based grinding media. For dolomite, see Chapter 13.

14.1.2 Sampling

Normal. The material needs to be dry as it tends to stick badly in the vial in the final grinding stages using a Tema mill. (The vial may need to be cleaned by grinding with wet clean sand.)

14.1.3 Drying

Normal. 'Dead-burnt' magnesites are resistant to the pick-up of water or carbon dioxide from the atmosphere, but other, less heavily ignited oxides may readily absorb both. Drying times need to be restricted, and if a knowledge of the carbonate content of the sample on receipt is important, special arrangements will need to be made for this.

14.1.4 Weighing

Normal. Lightly calcined oxides may need weighing rapidly to avoid errors from atmospheric absorption.

14.1.5 Loss on Ignition

Normal. The loss on ignition figure will be affected by the state of the dried sample. In the case of lightly calcined materials it is important to find an acceptable norm. A few natural materials can be very difficult to decompose completely at 1025°C, some carbon dioxide remaining. In these cases 1200°C may be needed.

14.1.6 Fusion

14.1.6.1 Flux

Lithium tetraborate.

14.1.6.2 Flux/Sample Ratio

10:1. 9.000 g ± 1 mg of flux:0.900 g ± 0.5 mg of ignited sample for a 35 mm diameter bead.

14.1.6.3 Fusion Procedure

Lithium tetraborate melts are comparatively viscous and may need swirling 2 or 3 times from the furnace.

14.1.7 Casting

Normal, but because of the increased viscosity it is important to make the transfer to the mould rapidly. The melt also has a tendency to devitrify, so providing an additional reason to cast it rapidly.

14.1.8 Constituents to be Determined

14.1.8.1 By XRF

Silica, alumina, ferric oxide, titania, lime, magnesia, soda, potash and phosphorus pentoxide. Zirconia is added to some grinding materials in per cent amounts. Chromium sesquioxide (some refractory magnesites may contain up to 5% of Cr_2O_3). Manganic oxide may be present.

14.1.8.2 By Other Methods

Soda, at low levels; sulphur trioxide; carbon, as an assessment of carbonaceous matter. The carbon in the carbonate, calculated as C, needs to be deducted from the 'carbon' figure determined by combustion. This may well mean, bearing in mind the possibly high carbonate content, that the carbon figure is unreliable; carbon dioxide from carbonates; boric oxide, usually at low levels up to 0.2% of B_2O_3. This determination is important where the material is used for refractory purposes.

14.1.9 Calibration Ranges

		%
Silica	SiO_2	0–5
Titania	TiO_2	0–5
Alumina	Al_2O_3	0–5
Ferric oxide	Fe_2O_3	0–10
Lime	CaO	0–5
Magnesia	MgO	80–100
Soda	Na_2O	0–5
Potash	K_2O	0–5
Chromium sesquioxide	Cr_2O_3	0–10
Manganic oxide	Mn_3O_4	0–5
Zirconia	ZrO_2	0–5

14.1.10 Reporting of Results

Normal. For natural carbonates, the loss on ignition figure is often used as the carbonate figure, as the errors in the latter determination at high levels are often greater than those entailed in using the loss figure.

14.2 MAGNESIUM SILICATES

14.2.1 Typical Materials

Talc, steatite, soapstone.

14.2.2 Sampling

Normal. Tends to stick in vial. (The vial may need to be cleaned by 'wet' grinding with clean sand.)

14.2.3 Drying

Normal.

14.2.4 Weighing

Normal.

14.2.5 Loss on Ignition

Normal.

14.2.6 Fusion

14.2.6.1 Flux

4 lithium metaborate:1 lithium tetraborate.

14.2.6.2 Flux/Sample Ratio

5:1. 7.500 g ± 1 mg of flux:1.500 g ± 0.5 mg of ignited sample for a 35 mm diameter bead.

14.2.6.3 Fusion Procedure

Normal.

14.2.7 Casting

Normal.

14.2.8 Constituents to be Determined

14.2.8.1 By XRF

Silica, alumina, ferric oxide, titania, lime, magnesia, soda and potash. Phosphorus pentoxide may be required.

14.2.8.2 By Other Methods

Carbon, as an assessment of carbonaceous matter; carbon dioxide from carbonates. The carbon in the carbonate, calculated as C, needs to be deducted from the 'carbon' figure determined by combustion.

Normally, in geochemical analysis the determination of ferrous oxide may be required.

14.2.9 Calibration Ranges

		%
Silica	SiO_2	50–100
Titania	TiO_2	0–5
Alumina	Al_2O_3	0–5

		%
Ferric oxide	Fe_2O_3	0–5
Lime	CaO	0–5
Magnesia	MgO	30–60
Soda	Na_2O	0–5
Potash	K_2O	0–5

14.2.10 Reporting of Results

Normal.

14.3 MAGNESIA/CHROME-BEARING MATERIALS

14.3.1 Typical Materials

Chrome ores, chrome-magnesite refractories (>20% of Cr_2O_3), magnesite-chrome refractories.

14.3.2 Sampling

Normal. If either cobalt or nickel is important, grinding in a tungsten carbide vial may be inadmissible.

14.3.3 Drying

Normal.

14.3.4 Weighing

Normal.

14.3.5 Loss on Ignition

Normal. Many of these materials contain spinels, some of which contain iron in the ferrous state (e.g. $FeO \cdot Cr_2O_3$). This ferrous iron is very difficult to oxidize. Ignition to constant weight is therefore necessary, but even so complete oxidation cannot be assured. This problem often reveals itself as a high analytical total when the determined total iron content is entered as being in the ferric state. This makes analytical totals within the normal limits difficult to achieve; a problem that is made no easier by the relative imprecision of the determination of magnesia. It is advisable, therefore, to confirm the accuracy of the results by carrying out duplicate analyses.

14.3.6 Fusion

14.3.6.1 Flux

Lithium metaborate and lithium tetraborate.

14.3.6.2 Flux/Sample Ratio

22.5:1. $5.000\,g \pm 1\,mg$ of lithium tetraborate $+ 4.000\,g \pm 1\,mg$ of lithium metaborate: $0.400\,g \pm 0.5\,mg$ of ignited sample for a 35 mm diameter bead.

The range of chromium sesquioxide contents that can be handled by this flux/sample ratio is 10–50% of Cr_2O_3. Below the lower figure the flux as for magnesites may be used. Above the higher figure dilution will be required; this is usually achieved by diluting with magnesia.

14.3.6.3 Fusion Procedure

Decomposition is slow, and is not apparently assisted by raising the temperature, although some analysts claim that 1400°C is advantageous. The thermodynamics of the equation whereby ferric oxide reduces to the ferrous state tend to favour reduction as the temperature is raised, possibly giving rise to increased problems caused by the ultimate reduction to metal and alloying with the metal of the dish. About 950°C appears optimal, but with oxidizing conditions being maintained throughout.

14.3.7 Casting

Normal, except that the bead may tend to stick. Addition of a crystal of lithium iodide or iodate to the dish immediately before casting helps.

14.3.8 Constituents to be Determined

14.3.8.1 By XRF

Silica, alumina, ferric oxide, titania, lime, magnesia, chromium sesquioxide and manganic oxide. Alkalis are not normally of significance if low in content; potash can be measured but the dilution of the fusion means that soda can only be estimated.

Oxides of lead, zinc, phosphorus, cobalt and nickel are commonly present at moderate trace levels.

14.3.8.2 By Other Methods

On the rare occasion when a soda figure is needed, this will have to be determined by alternative procedures. The difficulty in decomposing these sorts of materials makes this very difficult.

Occasionally in ceramic work, and normally for geochemical purposes, the determination of ferrous oxide may be required. Unfortunately, in this matrix the determination by chemical methods is particularly difficult and the results are doubtful.

14.3.9 Calibration Ranges

		%
Silica	SiO_2	0–10
Titania	TiO_2	0–5
Alumina	Al_2O_3	0–30
Ferric Oxide	Fe_2O_3	0–20
Lime	CaO	0–5
Magnesia	MgO	0–70
Soda	Na_2O	0–5
Potash	K_2O	0–5
Chromium sesquioxide	Cr_2O_3	0–50
Manganic oxide	Mn_3O_4	0–5

14.3.10 Reporting of Results

Normal. Chromium sesquioxide and manganic oxide are included in the total. After ignition, a gain in weight may be recorded because of the oxidation of ferrous iron to the ferric state. If this is so, the figure recorded should be deducted from the total (i.e. as a negative loss on ignition).

15 Procedures for Zircon-bearing Materials

15.1 TYPICAL MATERIALS

Zircon (zirconium silicate). Zircon used for ceramic purposes is usually pure, containing about 99% or more of zircon. However, some zircon derived from natural zircon sands (e.g. monazite) can be very impure, and may contain significant amounts of rare earths, tin, thorium, yttrium and uranium. If in doubt, it often pays to carry out a preliminary semi-quantitative analysis. The method below is intended for use with relatively pure zircons. Additional calibrations and interference investigations will deal satisfactorily with less pure materials. It should be noted that the presence of elements such as thorium or uranium can introduce hazards associated with radioactivity, the extent of which should be checked before proceeding. It has to be assumed that hafnia will be present.

Zircon-bearing refractories: kiln furniture. These may be included by extending calibrations.

Note: The decision to cover an extended range of calibration so as to include zircon-bearing refractories with the generic material would depend on the work load of each type. Slightly better accuracies can probably be obtained by separate calibration. It is even possible to include the mixed materials in the aluminosilicate range by adding the appropriate levels of zirconia, both in calibration and interference investigations. In any event, calibrations for minor constituents and their mutual interferences will be identical for both zircons and aluminosilicates.

15.2 SAMPLING

Normal. The materials are very hard and abrasive, and may thus be noticeably contaminated. Constituents other than zirconia and silica are usually at low levels. Care needs to be taken over the choice of grinding procedure.

15.3 DRYING

Normal.

15.4 WEIGHING

Normal.

15.5 LOSS ON IGNITION

Normal.

15.6 FUSION

15.6.1 Flux

4 lithium metaborate:1 lithium tetraborate.

15.6.2 Flux/Sample Ratio

5:1. 7.500 g ± 1 mg of flux:1.500 g ± 0.5 mg of ignited sample for a 35 mm diameter bead.

15.6.3 Fusion Procedure

Normal in terms of procedure. Fusion times may be longer, as zircon is more difficult to decompose than materials such as aluminosilicates.

15.7 CASTING

Normal.

15.8 CONSTITUENTS TO BE DETERMINED

15.8.1 By XRF

Silica, alumina, ferric oxide, titania, lime, magnesia, soda, potash, phosphorus pentoxide, zirconia and hafnia.

15.8.2 By Other Methods

Normally none. However, if soda and/or magnesia are required to below about 0.2% or 0.1% respectively, alternative techniques such as flame photometry or atomic absorption spectrometry will need to be used. Decomposition procedures will then need to be carefully considered, as zircon is not easily attacked by hydrofluoric acid.

15.9 CALIBRATION RANGES (ZIRCON)

		%
Silica	SiO_2	30–40
Titania	TiO_2	0–5
Alumina	Al_2O_3	0–5
Ferric oxide	Fe_2O_3	0–5
Lime	CaO	0–5
Magnesia	MgO	0–5
Soda	Na_2O	0–5
Potash	K_2O	0–5
Phosphorus pentoxide	P_2O_5	0–5
Zirconia	ZrO_2	55–70*
Hafnia	HfO_2	0–2

*For kiln furniture, etc., the calibration range is best extended down to 0%, or by using the aluminosilicate calibration as base.

15.10 REPORTING OF RESULTS

The normal eight oxides followed by zirconia, hafnia and loss on ignition.

16 Procedures for Various Oxides and Titanates

16.1 TYPICAL MATERIALS

Titania (rutile, anatase), ilmenite (FeO·TiO$_2$), zirconia (baddeleyite), zirconia (refractory and stabilized). Stabilized zirconia materials may have significant amounts of lime, magnesia and/or yttria added to achieve the desired properties.

Titanates: these may be described as barium, calcium, iron etc. depending on the other main constituent. The composition of these can be very variable. It is essential to carry out a semi-quantitative analysis first. In most cases the procedure described in this category may be used for bead preparation etc., but care needs to be taken if any of the constituents offer difficulties with regard to volatility, reaction with the alloy of the dish, reduction etc. The ultimate method of bead preparation and parameters used on the instrument may need to be modified to take account of variations in composition.

Iron oxide [haematite (Fe$_2$O$_3$), magnetite (Fe$_3$O$_4$), limonite (Fe$_2$O$_3$·xH$_2$O)], siderite (FeCO$_3$).

Manganese oxides: manganese dioxide [pyrolusite, psilomelane (hydrated) and precipitated], manganese sesquioxide (Mn$_2$O$_3$), manganite (hydrated), manganosic oxide (Mn$_3$O$_4$), hausmanite.

16.2 SAMPLING

Normal. Care needs to be taken to ensure that, if grinding is done in tungsten carbide and small amounts of specific oxides are required, contamination is not significant in respect of these oxides.

16.3 DRYING

Normal. It should be noted that, with a few of the materials in this group, those with oxides readily reduced or oxidized, care may need to be taken during drying to ensure no change in oxidation state. Also, hydrated minerals can, of course, lose water on drying, so that their treatment requires prior consideration to ensure that the required data is obtained.

16.4 WEIGHING

Normal.

16.5 LOSS ON IGNITION

Normal. Ignition should be carried out at temperatures and for times suitable for the combination of oxides present to ensure that they are in the accepted state of oxidation at completion. Repeat ignitions may be necessary.

16.6 FUSION

The data under this heading is dependent on the particular material being analysed. In some cases, where the overall basicity is low, some metaborate may be acceptable in the flux and a lower flux to sample ratio may be admissible.

16.6.1 Flux

Lithium tetraborate.

16.6.2 Flux/Sample Ratio

12:1. $9.600 \text{ g} \pm 1 \text{ mg}$ of flux : $0.800 \text{ g} \pm 0.5 \text{ mg}$ of ignited sample for a 35 mm diameter bead.

16.6.3 Fusion Procedure

Normal. Care needs to be taken to maintain oxidizing conditions throughout with types of sample containing easily reduced oxides.

16.7 CASTING

Normal.

16.8 CONSTITUENTS TO BE DETERMINED

16.8.1 By XRF

Silica, alumina, ferric oxide, titania, lime, magnesia, soda and potash. Any additional oxides needing to be determined will depend on the nature of the sample, and should include all those oxides, in addition to the above, that have been identified in determinable quantities in a semi-quantitative analysis. This statement includes those constituents normally determined by methods other than XRF.

When analysing titanates containing significant amounts of zinc, the determination of soda becomes effectively impossible; it will be necessary to refer to alternative techniques.

The determination of ferrous oxide may be required.

16.9 CALIBRATION RANGES

It is clearly not possible to give calibration ranges to cover such a wide variety of materials. The samples may best be treated as individual types, bracket standards being prepared, or, if any particular type is relatively frequently analysed, an appropriate calibration and interference study could be justified.

16.10 REPORTING OF RESULTS

Normal. Additional oxides are placed below the standard eight and with the loss on ignition usually added into the total. With some oxides the 'loss' may well be a 'gain' on ignition, in which case it is deducted from the total if the oxides are reported as usual in their stable (ignited) form. If, however, it is known, say, that a major constituent oxide is virtually all present in a different oxidation state, it may be more informative to report it so, re-calculating the loss to allow for the oxygen included. In these circumstances it is best to show the original figures as well as separate data, as the calculated analysis is based on certain assumptions which should also be stated.

17 Procedure for Glasses, Glazes and Frits

17.1 INTRODUCTION

Unlike most of the routine categories discussed in these chapters on procedures for specific types of materials, the general categories glasses, glazes and frits cover a wide range of compositions, not only in the range of contents of the oxides present but also in the range of oxides themselves. Thus, it is not possible to describe a simple universal procedure that will enable the analysis of the whole range of these materials to be handed successfully.

The most easily followed course is to describe the pattern of the nalysis required for materials of simple composition and to extend it to the more complex materials by indicating the necessary modifications called for by the presence of additional elements, such as are commonly found in regularly commercially used products. However, the list of additional elements cannot be comprehensive, as glass and glaze technology is changing continuously.

The range of products in this category may be thought to be different in differing establishments; also, it is possible that the range overlaps with products such as vitreous enamels or even welding fluxes. Once the category is extended to cover coloured glasses and glazes this overlap becomes more evident, although the complexity of vitreous enamels etc. is usually much greater. Types such as vitreous enamel are best analysed as 'one-offs', and this situation is dealt with in more detail in Chapter 19. This will not apply, of course, if a product or products of fixed composition are being analysed for routine control purposes, when special calibrations will be set up. In fact, many of the more complex glazes may be more economically analysed by a 'one-off' technique, always with the proviso that there is no continuing demand for the analysis of materials of similar composition. Clearly, these decisions have to be made in the light of the known or anticipated work-load of the laboratory.

These classes can include materials containing 12 or more oxides. This engenders a large number of calibration and α-coefficient standards in order to extend the elemental coverage. In addition, the computer software has to be able to cope with the sheer size of the matrices that have to be stored and the large numbers of line overlaps and α-coefficients. Many of the models used for corrections just will not work for large line overlap corrections combined with large α-corrections (not just coefficients; the correction being the product of the content of analysed oxide, the content of interfering oxide and the coefficient). The mathematical models used tend to be simplistic, and work well for the sorts of classes of materials dealt with up to this point. They can break down for complex matrices such as those in this section, and additional programming may be necessary.

17.2 TYPICAL MATERIALS

A change is made here in that, rather than itemizing specific material types, it seems more appropriate to itemize the additional elements or oxides (over and above the standard eight) that frequently occur in commerically used glasses, glazes and frits.

17.2.1 Colourless and White Materials

Boric oxide, fluorine, lithia, lead oxide, zinc oxide, zirconia (also introduces hafnia), barium oxide, strontia, stannic oxide, arsenious oxide (in glasses) and titania (deliberate addition).

17.2.2 Coloured Materials—Additional to 17.2.1

Ferric oxide (deliberate addition), manganic oxide, chromium sesquioxide, cobalt oxide, nickel oxide, copper oxide, cadmium oxide, cerium oxide, praseodymium oxide, uranium oxide, selenium (possibly as cadmium sulphoselenide, often mixed with barium sulphate).

High-temperature glazes may contain no additional elements or oxides than the standard eight to reduce the melting point below 1050°C. These may often be analysed by the aluminosilicate method.

17.3 SAMPLING

Normal. It is standard practice throughout the glass industry to grind only sufficient sample for immediate use. This is connected with the tendency of some ground glasses to react with moisture and carbon dioxide from the atmosphere.

17.4 DRYING

Routine. In a few cases it may be necessary to take account of the addition

PROCEDURE FOR GLASSES, GLAZES AND FRITS   233

to the mill batch of raw materials that can react with the atmosphere or that are significantly steam volatile. Colemanite, for example, if a component, will lose both water and boric oxide if dried at 110°C, unless previously fritted.

17.5 SEMI-QUANTITATIVE ANALYSIS

Unless the constituents are already known, in view of the many precautions that need to be taken in the presence of constituents other than the usual eight, a semi-quantitative analysis is best carried out. It is often possible, when analysing white or transparent glazes, to assume that, in addition to the eight, they may contain lead, zinc, zirconium (and therefore hafnium) and barium (and, therefore, possibly small amounts of strontium). If these are analysed for, together with lithium and boron, most colourless glazes and frits will yield satisfactory analyses. In glasses the presence of arsenic is common, and with white glasses (opal), zinc and fluorine. Vitreous enamels may safely be assumed to contain fluorine and boron. Most of the above glazes etc. will contain boron, and often small amounts of lithium.

It is unfortunate that the presence of boron and lithium cannot be ascertained by an XRF scan, and fluorine can be detected only with modern instruments. Their presence often has to be assumed, and lithium and boron usually need to be determined if only to prove their absence. The same, to some degree, applies to fluorine. A knowledge of the technology of the types of material in question has to be brought into play, determinations being carried out on those of these three light elements likely to be constituents of the particular type being analysed.

17.6 WEIGHING

Normal.

17.7 LOSS ON IGNITION

As most of the materials dealt with in this section are designed to melt, or at least soften, at temperatures below 1000°C, ignition cannot conveniently be performed in the same dish as that used for the fusion. The standard temperatuure for the loss on ignition in glass analysis is usually 550°C. Temperatures as low as 550°C will not necessarily suffice for many glazes, as, in their preparation, raw materials such as clay or limestone may be mixed with fritted material in the final milling stage. These components, if present, will not be completely decomposed at low temperatures, so that the figure for the loss on ignition will be low.

For this reason, the usual technique for the loss on ignition determination on glazes and frits is to weigh the portion of material into a disposable porcelain crucible and carry out the ignition on this. A separate corrected weight portion of the sample is used for fusion and subsequent analysis. Ignition is carried out usually over gas burners (provided that there is no overriding need to ensure oxidizing conditions), slowly raising the temperature until the sample melts or the full heat of the burner is reached and then heating for the appropriate time. Prolonged heating is undesirable, as the material may well contain easily volatilized components. Very occasionally the composition of the material may be such that reaction with the porcelain of the dish may be a possibility. Recourse then needs to be made to platinum or platinum alloy, taking care to ensure that the treatment will not cause damage to the ware. From the series of cautions given above, it will be realized that some thought needs to be given to the probable chemistry of the sample before proceeding.

17.7.1 Problems due to Additional Components

Fluorine

Reaction can occur, on heating, between the fluorine present in the sample and the porcelain dish or, more probably, silica or boric oxide in the sample itself. Any reaction with other components in the sample will result in volatilization of SiF_4 and/or BF_3. As the full amount of B_2O_3 and, probably, SiO_2 will be determined in separate portions of sample, the lost oxides will have been added into the total twice. However, experience has shown that, provided that heating is carried out as suggested and not prolonged, this error does not seem to be reflected in the analytical total.

17.8 FUSION

17.8.1 Flux

4 lithium metaborate:1 lithium tetraborate.

17.8.1 Flux/Sample Ratio

5:1. $7.500 \, g \pm 1 \, mg$ of flux: $1.500 \, g \pm 0.5 \, mg$ of ignited sample for a 35 mm diameter bead.

17.8.3 Fusion Procedure

The presence of heavy metals such as lead, zinc, tin, antimony and arsenic can result in severe damage to the fusion vessel. Great care should be exercised to ensure oxidizing conditions throughout, and the final fusion stage should be conducted at a temperature no higher than 1050°C.

17.8.3.1 *Problems due to Additional Components*

Fluorine:
Again the problem of reaction; the main risk is with the boric oxide in the flux. In the presence of much fluorine it is advantageous to cool and weigh the melt after fusion, quickly re-heat and cast. Allowance should then be made for melt weight.

Lead, zinc, arsenic, antimony, tin, cadmium
Care needs to be taken with each of these elements as, if reduced, they will tend to attack the metal of the dish, causing severe damage. Reduction is easy, particularly with the use of gas heating or elevated temperatures.

Lead, zinc, arsenic, antimony, cadmium
These elements can be volatile in varying degrees; care should be taken to ensure correct determinations. Duplicates may be desirable.

Stannic oxide, cobalt oxide, copper oxide
These oxides can only be brought into the melt in limited amounts. Dilutions may be necessary. Stannic oxide will not dissolve, whereas cobalt oxide is easily reduced and copper produces beads that are not easily cast as they tend to shatter. Even below the content limit they tend to stick to the dish. Lithium iodide or iodate is needed as a releasing agent, but can cause problems of bead curvature.

17.9 CASTING

Normal. As mentioned above, the presence of copper and cobalt necessitates the use of lithium iodide or iodate to release the bead; the same is true of chromium, if to a lesser extent.

17.10 CONSTITUENTS TO BE DETERMINED

17.10.1 By XRF

Silica, alumina, ferric oxide, titania, lime, magnesia, soda, potash
Additional elements have been adequately covered in the paragraphs above. Many simpler glazes, glasses and frits can be handled by extension of the routine silica/alumina calibrations and line overlap and α-correction matrices. Commonly found oxides in all three types can be calibrated for and the appropriate interferences evaluated. After this, analysis within the ranges offers no problems once the bead is prepared. (See section below on calculation.)

17.10.2 Constituents Determined by Other Methods

Boric oxide, lithia, fluorine, sulphur trioxide
Some SO_3 may be retained after the fusion and after the loss on ignition determination. The amounts may not be, of course, the same, giving rise to some difficulties in interpreting the results. The sulphur is often added to glazes as plaster, to improve their rheology, rather than as a part of the basic formulation.

Occasionally the determination of *ferrous oxide* may be required in the analysis of glasses.

17.11 CALIBRATION RANGES

In the following it is assumed that the analyst will already have calibrated for silica/alumina materials (Chapter 12); in any event these calibrations will be required in order to analyse the types of materials in this section.

Some of these calibrations will need to be extended and other calibrations added.

As must be appreciated from the foregoing, it is not possible to evaluate definite calibration ranges. Instead, an indication is given of likely ranges for each of the types glass, glaze and frit (Tables 17.1 and 17.2).

In addition to calibrations which would involve three or four binary standards, plus a zero, it is necessary to produce ternary standards to derive α-coefficients. For a matrix of n oxides, $(n-1)^2$ standards are needed. To allow for adequate checking, some (at least one each of these) are best prepared in duplicate. If n is much greater than 10, it is clear that the number of standards required is formidable. The analyst has therefore to give careful consideration to the potential amount of use of such standards before embarking on what may be a mammoth task.

Table 17.1. Additional calibration ranges (%) for colourless and white glazes and glasses additional to silica/alumina range materials

Constituent	Formula	Glass	Glaze	Borax frit	Lead Bisilicate frit
Titania	TiO_2	0–5	0–20	0–5	0–5
Lime	CaO	0–20	0–20	0–20	0–5
Magnesia	MgO	0–10	0–5	0–5	0–5
Soda	Na_2O	0–20	0–20	0–20	0–5
Potash	K_2O	0–20	0–20	0–5	0–5
Lead oxide	PbO	0–40	0–25	(0–5)	50–70
Zinc oxide	ZnO	0–10	0–15	–	–
Zirconia	ZrO_2	0–5	0–20	–	–
Hafnia	HfO_2	0–5	0–20	–	–
Tin oxide	SnO_2	0–5	0–10	–	–
Arsenious oxide	As_2O_3	0–5	–	–	–
Barium oxide	BaO	0–10	0–10	–	–
Strontia	SrO	0–5	0–5	–	–

Table 17.2. Additional calibration ranges for colouring oxides

Constituent	Formula	Calibration range (%)
Manganic oxide	Mn_3O_4	0–10
Chromium sesquioxide	Cr_2O_3	0–10
Cobalt oxide	CoO	0–10
Nickel oxide	NiO	0–10
Copper oxide	CuO	0–10
Cadmium oxide	CdO	0–10
Cerium oxide	CeO_2	0–10
Praseodymium oxide	Pr_6O_{11}	0–10
Uranium oxide	U_3O_8	0–10

17.12 CALCULATION

If lithia, boric oxide, fluorine or sulphur is present in the sample interference effects will occur due to their presence. In view of the fact that the flux contains vast amounts of both lithia and boric oxide, it may seem strange that the relatively small additional amounts present in the sample should need to be taken into consideration. This, however, is because calibration and inter-element corrections are based on a fixed sample/flux ratio. Both lithia and boric oxide reduce the mass absorption of a sample matrix which is predominantly silica and alumina.

17.13 REPORTING OF RESULTS

Normal; oxides other than the usual eight need to be included and added into the total.

If fluorine is a constituent, the necessary correction for the oxygen equivalent needs to be made.

18 Procedures for Reduced Materials: Carbides, Nitrides and Elements

18.1 INTRODUCTION

There are a range of reduced materials used as, or in the production of, refractories and engineering ceramics that are of growing importance. Ceramics now being utilized for engineering or electronics promise to be of vital importance in the future. While it is true that in many cases the information sought is that concerning the molecular species present, information about the purity of both the raw material and the finished products in terms of elemental composition is also important. The presence of undesirable (or even desirable) impurities and their quantities, even at trace element level, can radically affect the performance of many of these products. There is thus a rapidly increasing demand for the analysis of such materials. Elemental analysis is also often useful to help collate the information obtained from speciation analysis* and as a check on the mineral composition.

There is little doubt that the ultimate way forward in the characterization of most of the manufactured products within this range depends not only on a 'chemical' or elemental analysis but also on quantitative determinations of the *compounds* present. The latter ability is still, generally, in its infancy, the research work being very difficult and the general level of accuracy poor (in relation to analytical accuracies). It is almost impossible to check the accuracy of any results obtained, because of the lack of certified reference materials or even the limited availability of pure compounds to mix together to form synthetic standards. This latter course is in any event open to question, as a mixture hardly provides the same physical matrix as the much more intimate juxtaposition of the compounds in the actual material. Techniques such as XRD, IR and thermal methods, together with more recent structural techniques will need to be brought into play before it is possible to combine mineralogical analysis with elemental analysis to form a sound method of characterization.

---

*Speciation analysis is defined as the analysis of the material in terms of its molecular compounds.

A further category has been added to this section, ferro-alloys. Although these are not ceramics and are, in fact, of concern mainly to the steel industries, they are, nevertheless, of the same general chemical nature and can be handled by techniques similar to those for the other types of materials described.

It is not intended to describe in detail the experimental procedures for reduced materials, as has been done previously with other types of materials. General guidelines, suggested procedures and necessary precautions are given for each class of material, together with a number of other comments. There is no doubt that these types demand no little skill and much personal attention. Success can only be achieved with experience and care: the less the experience the greater the care. If in doubt, proceed slowly and carefully; failure to do so will almost certainly be at the price of a platinum/gold dish. Unless the number of such analyses is potentially high enough to justify the acquisition of the necessary expertise, it may well be more economical, in the long run, to buy in the analyses, thereby allowing the contracting laboratory to hazard its own platinum ware.

A 'reduced' material such as silicon carbide increases in weight (owing to oxidation) during fusion, and, as the XRF method assumes a correct ratio of sample to flux in the bead, this comparison could become fallacious. For example, if pure silicon carbide is being analysed, only 1.0 g of sample will be needed to give 1.5 g of oxidized 'sample', i.e. silica, in the bead. Thus, lower weights of sample are required for all reduced types of material to allow for conversions to the normal oxidation states (Si→SiO$_2$). Carbon and nitrogen are 'burned out' and are of no further concern. It may be assumed that, in the original material, elements such as iron, aluminium and titanium will normally be in a reduced (elemental) state, as may calcium and magnesium (possibly as carbides), although it is probable that the alkalis may still be accountable as oxides. Calcium, if reduced to the carbide, could have been subject to hydrolysis by atmospheric moisture before the analysis is undertaken. It is normal practice to report aluminium, iron and titanium in elemental form and the alkaline earths and alkalis as oxides. The silicon is reported in the manner appropriate to the type of material being analysed, possibly in more than one combination state.

18.2 TYPICAL MATERIALS

Silicon carbide, silicon nitride, silicon, boron carbide, boron nitride, sialons, ferro-alloys.

18.3 GENERAL PROCEDURE

18.3.1 Loss on Ignition

Losses on ignition are not often carried out on these materials, as it is usually very difficult to be certain of the chemical reactions taking place. It is very possible that part of the sample may be oxidized during the ignition, causing difficulties in interpreting the results. If the determination is carried out at all, special conditions may be necessary. An essential part of the analyst's work is to decide how to approach such a determination, bearing in mind the nature of the sample and the data required.

18.3.2 Fusion

In the following discussion, it is assumed that a 35 mm diameter bead is to be used, but similar considerations apply whatever the size of bead. When dealing with a sample of unknown purity it is necessary, first, to make a sensible assumption about the level of purity in terms of mineral composition, and to prepare a melt by weighing enough sample to yield an estimated 1.5 g of oxidized sample in the melt. The melt should be cooled and weighed, re-melted, cast and analysed. From the weight and the results, a reasonably accurate estimate may be made of the analysis to allow a further bead (or beads) to be made using near enough to the correct weight of sample to produce approximately 1.5 g of oxidized sample in the final melt. This avoids errors due to calculation back from the weighed melt and, in the case of the use of the top surface, from errors in curvature due to thicker or thinner beads than are normal. If the weight of the first 'trial' is very low, it may be advisable to prepare a further trial much nearer to the correct bead weight.

It is essential to carry out the analyses of reduced materials in duplicate. Apart from the other difficulties, it is very difficult to ensure that all the iron has been converted to, or retained in, the ferric state and not reduced and then alloyed with the platinum. Failure of the iron figures to replicate shows that this has happened at least once. It is often adequate with this type of analysis if the iron content is low and not of great significance to report the higher of the two figures, but there are occasions where this is not so and further samples will need to be fused. A second advantage of the use of two beads is that checks are available for all the other figures. Unless the probable analysis of the material is known, it is usually best to prepare a single bead first and only prepare the duplicate when the correct weight of sample is known.

The basic fusion technique is common to most reduced materials. There is always a need to guard, as far as possible, against the possibility of a triple contact between the reduced material, the flux and the platinum alloy. This is usually achieved by fusing a known weight of lithium tetraborate containing the weight of boric oxide normally contained in the 7.5 g of 4:1 flux. This is swirled so as to run over the sides and base of the dish and then allowed to cool, forming a protective coating. The amount of lithia in the tetraborate used is subtracted from the weight normally contained in 7.5 g of 4:1 flux and the remaining weight of lithia, in the form of lithium carbonate, is mixed with the sample in the bowl of the dish.

Prepare a weighed dish as described above with the lithium tetraborate layer, and with the sample and lithium carbonate mixed together intimately in the bowl formed by the solidified flux. Heat the dish over a small gas burner; start with a very low flame and gradually adjust the temperature until the lithium carbonate begins to react with the sample, while allowing the lithium tetraborate layer to remain solid. The fact that the reaction is very exothermic increases the need to proceed slowly, particularly with those samples where it is suspected that silicon 'metal' may be present in any quantity.

In the case of silicon carbide, ferrosilicon and possibly other ferro-alloys, the reaction is best initiated by using a second small burner from above. For some types of sample, this may need to be directed in turn on to the various areas of the mix. The reaction, in the case of ferrosilicon at least, can be sufficiently exothermic to be often self-sustaining once it has been initiated.

When the reaction appears to be virtually complete (the time may vary from 10 min for ferro-alloys to several hours for silicon carbide) raise the temperature of the burner to full heat and maintain this for some time. Place a lid partially on the dish and transfer it to a position over a large burner (e.g. Amal or Meker) with the flame giving about the same heat as the small burner, and again increase the temperature slowly to the maximum heat. When all the sample appears to have decomposed, transfer the dish and melt to a furnace at 1200°C for about 5 min. Then transfer the dish and contents to a desiccator. When cool, weigh the dish and melt to obtain the final weight of the melt. Cast the melt in the normal manner, but retain (in the fusion dish) all the glass not transferable to the casting mould.

18.4 SILICON-BEARING MATERIALS

Many of these materials, with the notable exceptions of the sialons and ferro-alloys, yield a final analysis appearing to be that of a high-silica material.

18.4.1 Completion of the bead preparation procedures

In order to check the purity of the sample, analyse the bead for the relevant oxides and, if the analytical total lies between 96 and 102%, retain the bead as suitable for the ultimate analysis. The melt remaining in the dish may then be discarded.

If this condition is not met, the flux/sample ratio is sufficiently in error for corrective action to be needed. This may be carried out in one of three ways, depending on the analytical total achieved. If it is too high then the sample is too concentrated in the bead, and if too low, too dilute.

(a) If the total exceeds the upper limit then further flux needs to be added, this flux being 4 lithium metaborate:1 lithium tetraborate. Fuse, in the original dish, the additional flux together with the cast bead and the traces of glass remaining after casting. When homogeneous, re-weigh the melt and cast a new bead.

Weight (g) of 4:1 flux to be added = $(0.09d)/(1 - d/100)$, where d = analytical total $- 100$, or $(m-9) \times 5$, where m is the weight of the melt.

This assumes the use of anhydrous flux; if the flux is generally used with a 'drying factor' the above weight should be adjusted accordingly.

(b) If the total is below the lower limit, then it is necessary to weigh more sample into the dish and re-fuse in the flux from the cast bead and that retained in the dish. Particular care needs to be taken during this fusion, as it involves oxidizing the additional sample without the safeguard of a protective lining. The technique is probably only applicable to less difficult materials within the range. The weight of additional sample may be calculated as follows.

Additional weight (g) of sample = $(9-m)W/(m-7.5)$, or $1.2eW/(10-e)$, where W = the weight of material used to prepare the original bead, m = the melt weight and $e = 100 -$ the analytical total.

Alternatively, it may be simpler and usually better to fuse a new bead using appropriate weights of flux and sample. This should be done if the deviation from the limits quoted above is large.

When fusing the duplicate bead for the analysis, it is normal to establish the correct parameters on the first bead and then to prepare a second bead with the correct weight of sample. This weight is calculated using the formula:

Sample weight (g) for duplicate bead = $1.5W/(m-7.5)$, where m and W are as above.

18.5 SILICON CARBIDE

In addition to the above information, the following specific data needs to be noted.

Silicon carbide is now a generic term covering an increasingly broad spectrum of materials containing a major content of silicon carbide. In times past it was possible to assume that a sample received under the heading 'silicon carbide' was either a 'pure' SiC or, more probably in the context of the refractories industries, a clay-bonded silicon carbide, i.e. silicon carbide granules bonded with a refractory clay. Now, however, this type is much less frequently met with, the most common type being self-bonded and/or containing species: silicon carbide, silicon nitride, silicon oxynitrides, silica, 'metallic' silicon and possibly oxides of silicon of lower valency state than silica, e.g. monoxide, to say nothing of the presence of 'free' carbon, possibly graphite. A recent paper by Julietti and Reeve[1] details a method of speciation analysis of such materials.

With the earlier types of silicon carbide refractories, it was possible to carry out an elemental analysis, e.g. using XRF, and by means of the analytical total and the various oxide contents to calculate a 'silicon carbide' figure. During decomposition, silicon carbide is burned out to produce silica, giving a change in weight, and it is this change in weight that can, in principle, be used to estimate the silicon carbide content.

$$SiC + 2O_2 \longrightarrow SiO_2 + CO_2$$
$$28 + 12 \longrightarrow 28 + 32$$
$$40 \longrightarrow 60$$

Thus, 1.0 g of pure silicon carbide will be converted to 1.5 g of silica, so that the obtained analytical 'total', as usually expressed, for a pure silicon carbide should be 150%. It therefore follows that each 1% of silicon carbide will produce an enhancement of the analytical total by 0.5%, and from this an estimate of the SiC content of this type of material may be made. Any SiO_2 remaining is recorded as such. It will be appreciated that errors in any determinations, including the loss on ignition, or failure to determine any component will produce errors in the analytical total and hence in the SiC and SiO_2 figures. On the other hand, errors by almost any other method using the so-called 'rational analysis' techniques are high, and the procedures almost prohibitively expensive to perform, so that in the past calculation from an elemental technique produced a reasonable compromise for the technologist. However, if more than one reduced species is present this form of calculation becomes inadmissible. Such an approach can be justified only if it is possible to determine by other techniques the individual contents of other molecular species, e.g. by determining nitrogen and silicon contents, etc. and using these figures to estimate the contents of the appropriate molecular species. Methods exist for these sorts of determinations, but all are dependent on assumptions as to the nature of the actual species present and an ability to determine each accurately. The former is open to doubt and the latter poses serious questions. Such comprehensive 'rational' analysis techniques are very expensive to develop, and can be totally invalidated by the addition of a further component to the system, an occurrence only too likely in such a rapidly developing field. Nevertheless, XRF has been used in the past for this purpose and is still used, the results often being extrapolated further than is justified. In the absence of any inexpensive and reliable alternative this is possibly understandable, but the imperfections and assumptions of the procedure need to be put on record.

Silicon carbide itself is very hard to decompose. It is the most difficult of the reduced materials in this respect. It is attacked very slowly by the flux, and will decompose only where there is triple contact between oxygen, sample and flux. The oxygen is necessary to carry out the oxidation to silica, and the alkaline flux to take up the silica formed as lithium silicate and, in fact, to trigger off the oxidation. Fluxes containing an oxidant, e.g. nitrate, have been recommended but they have their disadvantages. They tend to damage the platinum ware and, if either the sodium or potassium salt is used, the corresponding oxide becomes undeterminable; lithium nitrate is very hygroscopic and is therefore very difficult to handle in practice. The presence of 'metallic' silicon presents a severe danger to the ware, since if it comes into contact with the alloy at elevated temperature in the presence of flux it will produce a pinhole. Much silicon, particularly if unexpected, can result in a cut on the metal of the ware at the level of the contact between the melt and air just as though it had been caused by scissors.

For a pure silicon carbide, 1.0 g of sample is fused with 2.228 g of lithium carbonate after the dish has been lined with 6.600 g of lithium tetraborate.

Once the bead has been prepared using a weight of sample equivalent to 1.5 g of oxidized sample the analysis reduces to that of a high-silica material, but it is necessary to bear in mind that only 1.0 g of sample has been used rather than 1.5 g. Calculations may be made of 'silicon carbide' content if appropriate. If nitrogen or total and free carbon have been separately determined the silicon nitride content may be derived from the first of these figures and a carbide content may be based on the second. However, it will be seen that this assumes that the whole of the nitrogen is present as silicon nitride and not as oxynitride or even aluminium or iron nitrides. Similarly, it assumes that no 'free' carbon or graphite is present in the sample. Neither of these assumptions is likely to be strictly correct, except for a very limited portion of the full range of these materials as currently available. If any calculations of this sort can be made, a figure may arise for the 'silica' content, but all these figures need to be assessed with care and their value concluded bearing in mind how many of the possible contingencies have been evaluated and allowed for and what reliance can be placed on the various assumptions.

18.6 SILICON NITRIDE

The following specific information is in addition to that given above.

Silicon nitride (Si_3N_4) is finding increasing use in engineering ceramics. There is little doubt that, as manufacturing techniques develop and research continues into improved nitride-based ceramics, there will be a vast increase in the tonnage produced. Being of a highly technical nature, such materials are dependent on both the molecular and elemental analysis even more than the carbide.

Fusion is made using the technique outlined above, lining the dish with lithium tetraborate and decomposing the sample admixed with lithium carbonate. Decomposition is much more rapid and less dangerous to the platinum ware than with the carbide, and the whole proceeds much more smoothly. The weight of sample has to be adjusted, bearing in mind the increase in weight as a result of the oxidation of the Si_3N_4 to silica and the loss of nitrogen.

$$Si_3N_4 + 3O_2 \longrightarrow 3SiO_2 + 2N_2\uparrow*$$
$$(3 \times 28) + (4 \times 14) \longrightarrow 3 \times (28 + 32)$$
$$84 + 56 \longrightarrow 180$$
$$140 \longrightarrow 180$$

Thus 1.167 g of silicon nitride will produce an oxidized weight of 1.5 g.

The flux required is 6.600 g of lithium tetraborate and 2.228 g of carbonate.

Apart from the nitrogen figure, the oxygen content of the sample is another technological consideration and would be required in the complete analysis of such a material.

18.7 SILICON (METAL)

The following specific information is additional to that above.

Silicon is one of the basic raw materials in the manufacture of some of the products discussed in this section and its elemental purity is therefore of concern. Normal elemental analysis is thus of paramount importance, but the determination of oxygen content may also be required.

As has already been stated above, silicon, probably because of its extreme reactivity, can be fatal to platinum ware. Although it is possible to use the direct fusion technique, albeit with the greatest possible care, it is very desirable to use a totally different approach. One or more grams of the sample are decomposed using a mixture of nitric and hydrofluoric acids. CARE needs to be taken concerning the degree of heat being applied, as the reaction is highly exothermic and the liquid can easily boil over. It is important to raise the temperature slowly and progressively until it is evident that any possibility of violent reaction is over. Evaporation is continued to dryness and is followed by further evaporation(s) with nitric acid to remove fluoride. After ignition to decompose nitrates as far as possible, sufficient pure silica is added to bring the combined weight of the residue and silica to 1.5 g. Fusion and analysis are then completed as for a high-silica material. Obviously no estimate of the Si content itself can be made if this procedure is used, but this is normally of little importance.

If the direct fusion technique is used, the sample weight is 0.700 g and the fluxes are 6.600 g of tetraborate and 2.228 g of carbonate for a 35 mm diameter bead.

---

*The actual oxidation state of the nitrogen lost has not been ascertained and is of no significance in this context.

18.8 SIALONS

These are various combinations of alumina and silicon nitride of the type $Si_{6-z}Al_zN_{8-z}O_z$. The type promises to become very important in the field of engineering ceramics. The analytical procedure is very similar to that used for silicon nitride. It has to be remembered that the range of possible compositions is wide, so that unless data concerning the anticipated composition is available it is almost certain that a preliminary bead will need to be cast and 'analysed'. The weight of the melt also needs to be determined. After this, the weight of sample needed to provide an equivalent 1.5 g of oxidized sample is used for the melt so as to yield a 5:1 flux/sample dilution. In the melt, the sample will be equivalent to a silica/alumina range sample, probably an aluminosilicate or aluminous material. If the fusion fails to yield a satisfactory analytical total, see Section 18.4.1.

A starting sample weight of 1.200 g should be used, mixed with 2.228 g of lithium carbonate in the bowl of a dish lined with 6.600 g of lithium tetraborate.

18.9 FERRO-ALLOYS

There are a range of ferro-alloys and, although, as yet, not all these types have been attempted by the authors by the general sort of technique described above, there seems no good reason why suitable modifications should not enable satisfactory beads to be cast and analyses conducted. The major content of these materials is iron (often about 70% of Fe), so that this means the use of the equivalent of lithium tetraborate flux rather than the 4:1 flux for many of the types of ferro-alloys. Such analyses as have been carried out have given very good results in comparison with other methods and, in fact, the technique has been used for the analysis of certified reference materials. The weights given in Table 18.1 have been calculated from the composition of the relevant ferro-alloys. Those for any new types could be calculated by similar methods.

In order to achieve a final melt composition equating to the use of a tetraborate flux, it is necessary to replace the tetraborate coating layer of the dish with boric oxide. Table 18.1 gives details of weights for sample and fluxes for each of the alloys. Of these only ferrosilicon and ferroniobium have actually been analysed, but there is every reason to expect that the others will offer little greater difficulty if the suggested techniques are used. After fusion, ferrotitanium, ferromanganese, ferroboron, ferrotungsten (b) and ferromolybdenum (b) will be effectively a melt of one part of oxides to 12 of lithium tetraborate. Ferrosilicon is equivalent to one part of oxides to five of 4:1 flux: similarly ferrotungsten (a) and ferromolybdenum (a) should produce a melt of 1.5 g of oxides in 7.5 g of 4:1 flux. Ferrochromium, being so high

Table 18.1. Decomposition parameters for ferro-alloys

Ferro-alloy	Sample wt. (g)	LiCO$_3$ wt. (g)	B$_2$O$_3$ or Li$_2$B$_4$O$_7$ wt. (g)	Type of bead produced
Boron	0.6583	4.194	7.586 B	Fe$_2$O$_3$ 1:12 in T
Manganese	0.5734	4.194	7.904 B	Fe$_2$O$_3$/MnO 1:12 in T
Molybdenum (a)	1.0132	2.228	6.600 T	Fe$_2$O$_3$/MoO$_3$ 1:5 in 4:1 flux
Molybdenum (b)	0.5404	4.194	7.904 B	Fe$_2$O$_3$/MoO$_3$ 1:12 in T
Niobium	1.040	2.228	6.600 T	Fe$_2$O$_3$/Nb$_2$O$_5$ 1:12 in T
Silicon	0.770	2.228	6.600 T	Fe$_2$O$_3$/SiO$_2$ 1:5 in 4:1 flux
Titanium	0.5262	4.194	7.904 B	Fe$_2$O$_3$/TiO$_2$ 1:12 in T
Tungsten (a)*	1.2243	2.228	6.600 T	Fe$_2$O$_3$/WO$_3$ 1:5 in 4:1 flux
Tungsten (b)*	0.6530	4.194	7.904 B	Fe$_2$O$_3$/WO$_3$ 1:12 in T
Vanadium	0.4977	4.194	7.904 B	Fe$_2$O$_3$/V$_2$O$_5$ 1:12 in T

Abbreviations: B = boric oxide; T = lithium tetraborate.
*(a) and (b) are alternative procedures.

in chrome ($\simeq$ 75%), would prove very difficult. For this reason, it has been omitted from the table. It would almost certainly require a dilution technique, and discretion could prove to be the better part of valour.

18.10 BORON CARBIDE; BORON NITRIDE

These materials are readily decomposed and should be handled by a technique similar to that used for silicon nitride. Their reactivity demands considerable care in the earlier stages of the fusion, until the vigorous part of the decomposition has been completed. It has to be borne in mind that the oxidized material will be boric oxide rather than silica, so that the analyst is committed to the analysis of impurities in a sample of boric acid. This may well mean a separate set of calibrations and interference effects, including the effect of boric oxide on the other determinands. The objective of the fusion, therefore, is to convert the boron in the sample to boric oxide, and the lithium carbonate used for mixing and the lithium tetraborate used for lining the dish into 9 g of lithium tetraborate during the fusion procedure. The XRF instrument merely determines

Table 18.2. Weights of fluxes and samples for boron carbide and nitride

Sample type	Sample wt. (g)	$Li_2B_4O_7$ wt. (g)	wt. Li_2CO_3 (g)
Boron carbide	1.500	4.409	2.006
Boron nitride	1.500	5.307	3.932

the impurities in a 'sample' of lithium tetraborate. An alternative approach would be to choose sample weights and lithium tetraborate and lithium carbonate ratios to produce the equivalent of a sample containing 100% of B_2O_3 fused at a ratio of 1:5 in the 4:1 flux.

Weights required for the two types are given in Table 18.2.

REFERENCE

1 Julietti, R.J. and Reeve, B.C.E. (1991) *Br. Ceram. Res. J.*, **90**, 85–89.

19 Procedures for Samples of Unknown Composition

19.1 INTRODUCTION

Some glasses and glazes of complex composition could be included in this category, but, as these have been dealt with separately in Chapter 17, reference to that section may prove helpful.

Some classes of materials may be received for analysis, e.g. vitreous enamels, that cover a vast range of compositions. This applies even to the range of oxides they contain as well as to their relative amounts. Samples may also be received about which so little is known of their type or origin that their composition is, at the outset, mainly conjecture.

As mentioned in the section on glasses, glazes and frits, the method described for these can be extended to cover more diverse materials such as vitreous enamels, welding fluxes etc. There is potentially an overlap between coloured glasses and glazes, colours, welding fluxes, vitreous enamels and some more complex slags, and thus similar approaches need to be taken to their analyses. In some cases the sample may be of a type familiar to the analyst, so that the extent of the problem can be defined, in which case it may be possible to proceed with the quantitative work immediately. If the analyst understands the technology of the materials to be analysed, it is again probable that the actual work can be undertaken with more confidence and less preliminary work. Knowledge of the probable range of oxides and their general levels of content can be very helpful in avoiding semi-quantitative scans. If this knowledge extends to the components likely to have been used in the formulation of the material, disastrous mistakes can be avoided. For example, it may be important to know whether the boron in a glaze has been introduced as colemanite, together with knowing if it is in the raw or fritted state, since, if raw, the mineral will lose boron as well as water on heating.

An analyst who appreciates the technology of the materials for analysis will know when to anticipate the possible presence of fluorine, boron and lithium as well as the heavy metals, and possibly more importantly, when they will *not* be there (at least as deliberate additions). The accuracies required of the results and the ease (or otherwise) with which they can be achieved will also be known.

A glass technologist will know the range of compositions found in a glass of specified type. A ceramic analyst will appreciate the likely composition of glazes for various classes of pottery, and how they are likely to have been compounded, together with the additional oxides used for obtaining certain colours or effects. This latter knowledge will be applicable in large measure to vitreous enamels, together with the fact that fluorine is an almost invariable component. Such information can save considerable time and effort, and therefore cost. It is worth bearing in mind that experience in the analysis of these sorts of complex material is likely to be dearly bought, so that if the requirement is infrequent or a 'one-off', it may well be better and cheaper to seek the professional services of a laboratory conversant with the type of analysis required.

19.2 TYPICAL MATERIALS

Colourless glass: special types, e.g. optical; coloured glass, possibly selenium-bearing; coloured glaze (for these three classes see Chapter 17); colour (glass and pottery); vitreous enamel; welding flux; kiln or furnace deposits; slags: many types of slag may be better analysed on a routine basis, particularly if a regular requirement.

The same principles can be applied to a wide variety of other inorganic types. If the nature of the material and its constituents are known, it is possible that it may fit on to one of the many types of calibration ranges described in the foregoing sections, particularly if they can be made subject to slight and legitimate 'stretching'. Where the nature of the material is completely unknown the analyst may be able to follow the technique described below, but always with the precaution of bearing in mind that XRF fails to detect elements lighter than fluorine or sodium (depending on the instrument). Kiln deposits, for example, often contain ammonium salts, and some slags or used blast furnace refractories may contain cyanide. Both of these will escape unless recourse is made to other, probably chemical, techniques. Equally obviously, any organic components will not be recognized, so that XRF has to be used with care and intelligence and should not be regarded as a universal panacea. Its limitations must always be borne in mind when analysing 'unknown' samples. If the circumstances can give rise to any possibility of doubt, it is usually advisable to check for the presence of fluorine, lithium and boron, and, where appropriate, ammonium and cyanide. The ammonium is usually present as ammonium sulphates and there is often free sulphuric acid (!) as well; there may also be, less frequently, chloride. Even here, however, the purpose of the analysis may be such that the presence and nature of volatile components may be of no consequence, as, for example, when the interest lies in the analysis of the residual material after ignition. Nevertheless, in the types of material likely to be handled by these techniques it is most probable that a complete breakdown of all the constituents is necessary, with the possible exception of organic components. These are, however, a separate issue and would probably be handled by a different analytical unit.

19.3 BASIC PROCEDURE

19.3.1 Sampling

Normal. A few of these sorts of samples can give rise to problems during grinding because they are hygroscopic (or even deliquescent) or they tend to decompose as a result of the heat generated. Difficulty will also arise if they contain tarry matter or, possibly, resins; these will almost certainly result in 'smearing' in the vial and, in the extreme, the complete prevention of grinding, at least by mechanical means. It is also possible that some of these types of materials could damage the grinding vials. It may prove necessary to carry out the analysis on the material after removal of the offending constituent(s) by ignition, dissolution etc. If this course is adopted, the nature of the material separated will need to be ascertained and possibly analysed, then the two analyses combined.

19.3.2 Semi-quantitative Analysis

In order to achieve a 'full' qualitative or semi-quantitative analysis, a sequential or simultaneous/sequential instrument is clearly necessary. Otherwise the analysis is restricted to those elements covered by the available channels of the spectrometer, which may well be of little service.

Once the sample has been reduced to powder form, an identification of the composition of the sample can be carried out. Obviously, damp materials, particularly if they are acid in nature, could do serious damage to a spectrometer and should NOT be introduced.

A semi-quantitative analysis may be carried out by the simple expedient of rubbing some of the sample powder into an acid-washed filter paper suitable for chemical methods of quantitative analysis (such as Whatman 42) cut, if necessary, into a circle of diameter to fit conveniently into a sample holder of the spectrometer. After gently tapping to remove surplus powder, a hair lacquer is sprayed on to the reverse side of the paper. This has the effect of binding the powder to the paper. It is, of course, necessary to check that the chosen lacquer does not itself give rise to any positive readings capable of being mistaken for possible constituents. Lighter elements can be scanned using limited ranges on the TlAP, Ge, and PE crystals, while most of the remainder can be covered by the use of a LiF 200 crystal. It may be possible with a modern instrument to identify the presence of fluorine at sufficiently low levels. The output of the instrument is often obtained in the form of a trace relating wavelength to magnitude of fluorescent response. This is compared with a previously prepared trace showing the positions of peaks (wavelengths) for the various elemental lines, by which comparison the elements present may be identified*. The height of the peak also enables estimates to be made of the respective contents. After some experience, it is surprising how reliable these estimates can be, but there is of course always the exception. This occurs most frequently with unusual matrices that may give rise to radical changes of mass-absorption coefficients. Fortunately the use of a very thin film of material almost eliminates problems of this sort, as the layer of material through which the emerging fluorescent X-rays have to pass is so slight. It is obviously conceivable that such identification and quantification could be greatly assisted by suitable computer programs, but these are still fairly uncommon and expensive. If a simultaneous-type instrument is available it is advantageous to put the paper disc into this instrument also. Not only does this produce actual counts that are easier to interpret as content levels (and usually more accurate), but it also saves time in interpreting the chart, as there is no need to seek the data concerning those elements available on the simultaneous instrument. There are, however, problems with simultaneous instruments, as backgrounds are difficult to estimate.

In addition, it is necessary to check the sample for the presence of those elements not detectable by XRF: fluorine (in the case of older spectrometers), boron and lithium. This is not necessary for all samples. If there is enough information on the type or origin of the sample, the presence or absence of each one of these may often be inferred. Otherwise it will be necessary to use alternative techniques to identify the presence or otherwise of these elements. They are all of the type for which it is usually simpler (and less expensive in the long run) to make actual determinations immediately.

The use of this semi-quantitative information is described as required in the sections below.

19.3.3 Drying

Normal. Care needs to be exercised with some types of sample (see comment under Sampling). If there is any doubt about whether the sample should be dried or not, the technique of using the sample in the air-dried state should be followed. If this is done, *all* portions of the sample should be weighed in one session, together with a portion on which the loss of weight on drying may be obtained. If the dried material can be regarded as a stable state, it will form a satisfactory basis to which to relate the analytical results, but if this was so, in most cases it would be possible to dry the sample to this state. It is not difficult

---

*'Spectromel' powder, supplied by Johnson-Matthey for spectrographic use, has been found to be a useful source of many elements for such a trace.

to visualize compositions within this very wide range that would give rise to problems of instability to atmosphere. These would not give a stable 'platform' from which the results could be calculated with any real confidence. In these circumstances it is only possible to use the best compromise, bearing in mind the purpose of the analysis and the information sought. It may be extremely difficult, if not impossible, to establish a stable base from which the analysis can be calculated, except possibly in the 'as received' state. However, time will have elapsed since the sample left the bulk it is intended to represent. Both sample and bulk will have since been subjected to different environments. Relating the analysis arrived at to the bulk whose analysis is required has to be carefully considered and techniques must be thought out for the results to be properly related. This is a difficult area of analysis, and the ability to make the right decisions can only come with experience.

19.3.4 Weighing

Normal, provided that none of the problems with instability to atmosphere apply. Where problems arise, specific solutions need to be applied.

19.4 VARIANTS IN PROCEDURE REQUIRED BY SAMPLE TYPE

GLASSES AND GLAZES

See also Chapter 17.

VITREOUS ENAMELS, ETC.

Once there is an intention to make coloured products, the range of elements to be sought is greatly increased. Each colour requires the use of a particular oxide (or, more probably, a combination of oxides) so that, although the potential range of permutations is vast, the analyst experienced in this field will generally be able to restrict the possibilities by consideration of the colour produced and its particular purpose. It is probable that black is the most difficult in this respect, as it is often produced by the admixture of several oxides, each producing a dark colour. It is not the intention here to enter into this technology; in any event, materials of this complexity should always be subjected to a preliminary semi-quantitative analysis except in the few instances where the general composition is already known.

Most of the oxides already encountered in colourless glasses and glazes may be anticipated in *coloured glazes* or *colours*. Fluorine is likely to be present, and it is not advisable to assume its absence even if the apparent analytical total is within tolerance limits. Any change in analytical total is the difference between the fluorine present and the equivalent oxygen (in round terms, about one half the fluorine content); the position is further complicated by the fact that some of the oxides may be present in a different oxidation state (even after the loss on ignition) than is assumed in the total. Oxides such as cobalt oxide are very difficult to convert to a definite oxidation state. In addition, there is always the uncertainty surrounding the oxidation state of reduced oxides in compounds such as silicates and spinels. The presence of lithia and boric oxide should be assumed, and their contents determined by suitable methods.

Vitreous enamels and *welding fluxes* will almost certainly contain analytically significant amounts of fluorine, and the former will need to be checked for lithia and boric oxide. Welding fluxes will usually contain significant amounts of lime, and because they are not specifically intended as colours the types and amounts of transition metal oxides are less predictable.

Colouring oxides or elements frequently encountered include the following.

Antimony pentoxide: tends to be volatile and damaging to platinum ware.
Arsenic pentoxide: tends to be volatile and damaging to platinum ware.
Ceric oxide.
Chromium sesquioxide (sometimes chromate): restricted range ($< \simeq 10\%$), lithium iodide or iodate may be needed.
Cobalt oxide (Co_3O_4): restricted range ($< \simeq 10\%$), alloys with platinum ware, oxidizing conditions essential, lithium iodide or iodate may be needed.
Cupric oxide: restricted range ($< 3\%$), alloys with platinum ware and causes sticking of the melt, oxidizing conditions essential, lithium iodide or iodate may be needed.
Ferric oxide: watch for reduction.
Manganese dioxide: lithium iodide or iodate may be needed.
Molybdenum trioxide: tends to be volatile.
Nickel oxide: lithium iodide may be needed, alloys with platinum ware, oxidizing conditions essential, but may still be problematical.
Praseodymium oxide.
Selenium (both as element and as cadmium selenide or sulphoselenide): extremely volatile; cannot therefore be determined by fused, cast bead technique.
Tungstic oxide: alloys with platinum, lithium iodide or iodate may be needed.
Uranium oxides.
Vanadium pentoxide: tends to be volatile.

Selenium is very difficult to determine in these materials by almost any technique. This is because decomposition can be achieved only by fusion or hydrofluoric acid treatment. The first will tend to cause loss of selenium because of its ready volatility and the second will involve the potential loss of gaseous hydrogen selenide (H_2Se). This tends to pose a serious problem in that the results obtained are very difficult to substantiate. This has been discussed previously (Chapter 3). This situation may not be applicable to the new encapsulated cadmium sulphoselenide colours, which are designed specifically to overcome the loss of cadmium.

19.4.1 Fusion

19.4.1.1 Flux

4 lithium metaborate:1 lithium tetraborate.
If the material contains an abnormally high level of total metallic oxides it may be necessary to use lithium tetraborate as flux.

19.4.1.2 Flux/Sample Ratio

Usually 5:1. 7.500 g ± 1 mg of flux:1.500 g ± 0.5 mg of ignited sample (equivalent) for a 35 mm diameter bead.
See also Chapter 17 concerning white, opaque glasses.

19.4.1.3 Fusion Procedure

The temperature of fusion will probably have to be restricted to 1050°C, depending on the constituents identified as being present. Precautions will need to be taken, as may be appropriate, based on the constituents.

19.4.2 Casting

Normal. Lithium iodide or iodate is likely to prove necessary to assist in the removal of the bead from the casting mould without its cracking.

19.4.3 Constituents to be Determined

By XRF

Silica, alumina, ferric oxide, titania, lime, magnesia, soda, potash. Additional constituents are as identified by the semi-quantitative analysis. Small amounts of constituents are not normally required; specific problems may involve seeking the presence of, and quantification of, specific elements.

By Other Methods

Boric oxide, fluorine, lithia; sulphur trioxide, carbon and carbon dioxide may be required.

19.4.4 Calibration Ranges

Unless such samples are regularly analysed, the use of normal calibrations is often the more complex way of proceeding with materials as complicated as these. It is often more appropriate to prepare a bead of the sample, assess the levels of content of each oxide and then prepare 'bracket standards' whose composition lies on each side of that anticipated. If the original assessment has been sufficiently good, it is probable that the results so obtained will suffice; normally, further bracket standards may be necessary.

The alternative approach via already existing calibrations and interferences may well be better if the range of available knowledge and standards is wide enough to cover the desired analysis, at worst without undue extrapolation. This approach is basically the simplest, if appropriate, but it could involve a considerable number of standards and much calibration effort.

On the rare occasion where work of the highest possible standard is needed, confirmation of the accuracy of the results may be achieved by preparing a bead of a synthetic sample made to the determined composition from pure oxides, etc. The accuracy of the results on this bead can usually be related directly to those on the sample being analysed. If significant discrepancies occur between the known analysis of the as-prepared and the determined figures, investigation of the causes will need to be made, as it is most likely that similar errors will apply to the sample under analysis. It must be pointed out, however, that the preparation of standards containing a number of oxides is a slow and laborious process, and is also prone to error, as it is very difficult to ensure that each oxide has been correctly weighed out, particularly as some of those oxides already weighed out could be absorbing atmospheric moisture or carbon dioxide during the weighing period.

19.4.5 Calculation of Results

The normal principles apply, but with additional complexity because of the increased number of components, and in correcting 'apparent' figures to 'true' figures using the bracket standards.

19.4.6 Reporting of Results

Oxides should normally be reported in their stable 1025°C ignited (and cooled) form except where there is reason to suspect unusual oxidation states in the sample.

SLAGS

The same procedures apply as have been discussed above. Most slags are derived from iron or steel making; reference to the composition of suitable CRMs will give a guide to the likely composition if the source of the sample is known. Semi-quantitative checks may be advisable. High contents of lime will probably be present, and fluorine may also be present, either at low levels of content (basic slags) or in percentage quantities depending on the process from which the slag has been derived. Iron may well be present as metal or in both oxidation states, ferrous and ferric. The free metal demands considerable care during the fusion to prevent alloying with the metal of the dish. Very significant amounts of phosphorus pentoxide, manganous oxide and chromium sesquioxide may also be present as well as the normal eight oxides.

Slags from the manufacture of non-ferrous metals may need to be analysed; if the analyst is familiar with the technology of the process the approximate composition will be known, otherwise it is highly desirable to carry out a full semi-quantitative analysis before proceeding further.

Where actual metal is present, it is unlikely that a simple loss on ignition will be of service. Apart from any other considerations that have been discussed elsewhere in this book, it is almost impossible to convert all the metal in the granules to the oxide. Once a surface coating of oxide has formed, the remainder of the granule becomes increasingly protected from the atmosphere, thereby preventing further oxidation. The presence of metal in any significant amounts can cause difficulty in grinding a sample, as it will tend to smear rather than grind. This is, again, a situation where the needs of the analysis must determine the course of action.

Laboratories called upon to analyse slags on a regular basis will obviously set up the appropriate calibrations.

KILN DEPOSITS

It is very infrequently that more than an approximate semi-quantitative analysis is required of such materials. After all, the acquisition of the sample is rarely specific; the usual requirement is to provide the technologist with an indication of what is volatilizing and then condensing in the kiln. By definition, most of the components are volatile, and it is essential to seek non-XRF-determinable components. These may well include ammonium, sulphur (combined as ammonium sulphates and as free sulphuric acid), borate, lithia and fluorine. In addition, such volatiles as lead, arsenic, antimony, zinc, cadmium, chromium, vanadium, nickel, germanium, sodium and potassium are often present. The list is not comprehensive and consideration needs to be given to all elements present in the kiln from whatever source. There will also be dust carried from almost anywhere to be deposited at odd points in the kiln. The ware itself, together with the glaze, and the setting it is placed on, and even the heating elements (if an electric kiln is involved) and the metal of the gas burners, to say nothing of the fuel itself, can all contribute to the nature of the final deposit.

The increasing use of organic materials in the manufacturing processes introduces nitrogen from which ammonium can appear. In the unlikely event that coal has been used as the fuel, all the inorganic components of the coal will need to be considered, possibly even if it is a muffle kiln.

If the analysis has to be completed reasonably accurately, the same general approach as has been described above will need to be used. Difficulties will arise from the volatility of many of the components and the need to make a number of non-XRF determinations. If any of these non-XRF components is likely to be present in the bead it may be necessary to make interference corrections for their presence.

It will be seen from the above that such an analysis will be very expensive, so that it will only be called for very infrequently when circumstances make it imperative.

Appendix I
Loss on Ignition Techniques

The following table gives the requirements for loss on ignition determinations for a range of types of samples. Six items of information are given for each type as follows:—

Drying temperature and conditions.
Number of samples usually used for routine work. Where this is not 1, duplicates will be normal, and the degree of replication of the results needs to be considered for various reasons. For referee work, duplicates, at least, would be the norm.
Nature of the ignition vessel.
Method of raising the temperature to the ultimate; where 'none' is specified, the sample may be introduced immediately to the ignition temperature.
Final temperature.
Desiccant required.
Variants may need reference to the particular (Chapters 12–19).

ABBREVIATIONS

The first entry given is for 'Routine' samples, and shows the conditions used for most samples. Where these conditions apply the entry is 'R'.

Preheat

Pt, platinum crucible; P, porcelain crucible; LSB, heat sample first on a low flame of a small bunsen burner; O/N, heat overnight.

Ignition Temperature

TCW, Requires re-heating to constant weight.

Desiccant

$CaCl_2$, calcium chloride; CSB, Carbosorb or similar absorbent for carbon dioxide.

Material type	Drying temp. (°C)	Number of replicates	Nature of vessel	Pre-heat conditions	Temp. of ignition	Desiccant
Routine	110	1	Pt/Au 5%	Slow to 1025°C	1025	Silica gel
Albite[a]	R	1	R	R	R	R
Alumina	R	1	R	R	R	R
Aluminosilicates	R	1	R	R	R	R
Aluminous materials	R	1	R	R	R	R
Amosite	See Note 1					
Anatase	See Titania					
Andalusite	R	1	R	R	R	R
Anorthite[a]	R	1	R	R	R	R
Asbestos	See Note 1					
Attapulgite	R	1	R	R	R	R
Baddelyite	R	1	R	R	R	R
Ball clay	R	1	R	R	R	R
Barium carbonate	R	1	R	R	R	R
Barytes	R	1	R	R	R	R
Bauxite	R	1	R	R	R	R
Bone ash	R	1	R	R	R or 1200	R
Bone china body	R	1	R	R	R	R
Boron carbide	R	No loss on ignition carried out				
Boron nitride	R	No loss on ignition carried out				
Brucite	R	1	R	R	R	$CaCl_2$ + CSB
Calcite	R	See Limestone				
Calcium borosilicate	See Colemanite					
Calcium carbonate	R	See Limestone				
Calcium fluoride	R	1	R	R	500	R
Calcium phosphates	R	1	R	R	R or 1200	R
Calcium silicates	R	1	R	R	R	R
Carbon	R	2	Pt	LSB, 600°C O/N	R + TCW	R
Cement	R	1	R	R	R	R
Cement + organic	R	1	R	LSB 20 min + R	R	R
Chalk	R	See Limestone				
Chert	R	1	R	R	R	R
China clay	R	1	R	R	R	R
Chromite	R	1	R	R	R	R
Chrysotile	See Note 1					
Clay	R	1	R	R	R	R
Coal	R	2	Pt	LSB, 600°C O/N	R + TCW	R
Cobalt aluminate	R	2	R	None	700	R
Cobalt oxide	R	2	R	None	700	R

APPENDIX I LOSS ON IGNITION TECHNIQUES

Table *(continued)*

Material type	Drying temp. (°C)	Number of replicates	Nature of vessel	Pre-heat conditions	Temp. of ignition	Desiccant
Routine	110	1	Pt/Au 5%	Slow to 1025°C	1025	Silica gel
Colemanite	*No drying or heating, sample reported on 'as received' basis*					
Colours	R	*See Note 2*				
Corundum	R	1	R	R	R	R
Cristobalite	R	1	R	R	R	R
Crocidolite	*See Note 1*					
Diatomite	R	1	R	R	R	R
Dicalcium phosphate	R	1	R	R	R + 1200	R
Dolomite	R	1	R	R	R	$CaCl_2$ + CSB
Feldspar[a]	R	1	R	R	R	R
Ferromanganese	None	*Normally no loss on ignition*				
Ferrosilicon	None	*Normally no loss on ignition*				
Ferrotungsten	None	*Normally no loss on ignition*				
Ferrovanadium	None	*Normally no loss on ignition*				
Fireclay	R	1	R	R	R	R
Fluorspar	R	1	R	R	500	R
Flint	R	1	R	R	R	R
Fluxes	R	*See Note 2*				
Gibbsite	R	1	R	R	R	R
Glasses	R	1	P	*See Note 2*		
Glazes	R	1	P	*See Note 2*		
Graphite or graphite-bearing	R	2	Pt	LSB + 600°C O/N	R + TCW	R
Graphitic SiC	R	2	Pt	LSB + 600°C O/N	800°C + TCW	R
Gypsum	None	2	R	None	600	$CaCl_2$ + CSB
Haematite	R	1	R	R	1025 + TCW	R
Illite	R	1	R	R	R	R
Iron oxides	R	1	R	R	1025 + TCW	R
Iron silicates	R	1	R	R	1025 + TCW	R
Kaolinite	R	1	R	R	R	R
Kyanite	R	1	R	R	R	R
Lead bisilicate	R	1	P	*See Note 2*		R
Lead borosilicates	R	1	P	*See Note 2*		R
Lepidolite[a]	R	1	R	R	R	
Limestone, lime	R	1	R	R	R	$CaCl_2$ + CSB

continued overleaf

Table *(continued)*

Material type	Drying temp. (°C)	Number of replicates	Nature of vessel	Pre-heat conditions	Temp. of ignition	Desiccant
Routine	110	1	Pt/Au 5%	Slow to 1025°C	1025	Silica gel
Magnesia	R	1	R	R	R	CaCl$_2$ + CSB
Magnesite, raw	R	1	R	R	R	CaCl$_2$ + CSB
Magnesium carbonate	R	1	R	R	R	CaCl$_2$ + CSB
Magnesite, dead-burnt	R	1	R	R	R	
Magnetite	R	1	R	R	1025 + TCW	R
Manganese oxides	R	1	R	R	1025 + TCW	R
Marble	R	*See Limestone*				
Marl	R	1	R	R	R	R
Mica[a]	R	1	R	R	R	R
Manganese silicate	R	1	R	R	1025 + TCW	R
Molochite	R	1	R	R	R	R
Molybdenum oxide	R	1	R	None	700	R
Montmorillonite[b]	*See footnote*	1	R	R	R	R
Mullite	R	1	R	R	R	R
Muscovite[a]	R	1	R	R	R	R
Nepheline syenite	R	1	R	R	R	R
Nickel oxide	R	1	R	R	700	R
Olivine	R	1	R	R	R	R
Orthoclase[a]	R	1	R	R	R	R
Periclase	R	1	R	R	R	CaCl$_2$ + CSB
Petalite[a]	R	1	R	R	R	R
Plaster	None	2	R	None	600	CaCl$_2$ + CSB
Quartz	R	1	R	R	R	R
Quartzite	R	1	R	R	R	R
Rutile	R	1	R	R	R	
Sand	R	1	R	R	R	R
Serpentine	R	1	R	R	R	R
Silica	R	1	R	R	R	R
Siliceous gunning mix	R	2	Pt	LSB + R	R	R
Silicon	*No loss on ignition carried out*					
Silicon carbide	R	2	R	None	750	R
Silicon nitride	R	*Normally no loss on ignition*				
Sillimanite	R	1	R	R	R	R
Slags[a]	R	1	R	R	R	R
Soapstone[d]	R	1	R	R	R	R
Spinel	R	1	R	R	R	R

APPENDIX I LOSS ON IGNITION TECHNIQUES

Table *(continued)*

Material type	Drying temp. (°C)	Number of replicates	Nature of vessel	Pre-heat conditions	Temp. of ignition	Desiccant
Routine	110	1	Pt/Au 5%	Slow to 1025°C	1025	Silica gel
Stannic oxide	R	2	Pt	None	700	R
Steatite	R	1	R	R	R	R
Talc[d]	R	1	R	R	R	R
Titania	R	1	R	R	R	R
Vermiculite	R	1	R	R	R	R
Vitreous enamels	*Handle as may be demanded by the constituent elements and oxides*					
Welding fluxes[c]	*Handle as may be demanded by the constituent elements and oxides*					
Whiting	R	*See limestone*				
Witherite	R	1	R	R	R	R
Wollastonite	R	1	R	R	R	R
Zeolites	R	1	R	R	R	$CaCl_2$ + CSB
Zinc oxide	R	1	R	R	R	R
Zircon	R	1	R	R	R	R
Zirconia	R	1	R	R	R	R

[a]Sample may sinter sufficiently to prevent the same portion being used for fusion etc.
[b]Care needs to be taken in drying; consult information in the text.
[c]The procedure for the loss on ignition determination may need to be modified to allow for volatile elements being present. Fluorine and sulphur oxides may be expected, but volatile metallic elements may also be present.
[d]Some of these materials may contain significant amounts of dolomite. Care then needs to be taken to avoid CO_2 absorption.

Note 1: Asbestos
Samples of asbestos or containing asbestos must be handled with great care to prevent any dust from it being inhaled. Gloves and a respirator should be worn. Any handling (including weighing) should be carried out in a fume cupboard on a large sheet of clean paper. A wash bottle should be available to damp down any spillage. The debris from any handling should be disposed of safely, as befits the nature of the material. The dangerous nature of the material is destroyed by heat, so that after ignition further precautions are not necessary. The complexities of handling asbestos may well dictate the use of an accredited laboratory with the necessary expertise.

Note 2: Colours, colemanite, fluxes, glasses, glazes, lead bisilicate, lead borosilicates
These types of samples will tend to melt well below 1025°C. The ignition should be carried out in a porcelain crucible being discarded after completion of the determination. The temperature should be raised very slowly to a dull red heat, attempting to retain the sample mix just below softening point for some minutes. For glasses a temperature of 550°C is used for the loss on ignition, by definition. Other materials should have the heating continued, gradually raising the temperature even to the full heat of a Meker burner, if appropriate. The correct top temperature may depend on the nature of the material and the information required from the analysis. These points are discussed in detail in appropriate places in the general text and under specific materials (Chapter 17). The remainder of the analysis will need to be conducted on a fresh portion of the sample using a weight that allows for the determined loss on ignition.

Appendix II
Specific Fusion Techniques

The following table gives details of the fusion techniques to be used for specific types of material. The coverage is not universal, but sufficient material types are included for the analyst to identify a similar type in most instances.

The information given includes the choice of flux, the flux/sample ratio as weights of ignited sample and 'dried' flux, the procedure for fusion, the casting temperature, an indication as to whether or not the sample requires a releasing agent such as lithium iodide (or iodate), and finally whether it is necessary to weigh the melt to determine any possible losses in weight during fusion (as distinct from losses during ignition). The routine process is shown in the first line of each page and, when this is the correct approach for the material in the current line of the table for that part of the data, that column is marked with an 'R'.

ABBREVIATIONS

Flux

Four parts by weight of lithium metaborate to one of tetraborate: 4/1; lithium metaborate: M; lithium tetraborate: T; lithium carbonate: L; boric oxide: B; mixtures of fluxes other than 4/1 are given as T/B, i.e. tetraborate with boric oxide, etc., the respective weights being given in the next column in an identical order. All weights are given in grams.

Others

RE, routine except an extended fusion at 1200°C is needed to ensure complete decomposition.
CL, the dried sample is weighed out, corrected for its loss on ignition.
Cooling the melt in the dish, weighing, remelting and casting: yes: Y, no: N.
GB, heat on gas burner progressively raising the temperature to full heat, swirling periodically to ensure complete mixing. This is normally combined with the next abbreviation as the sample is most often transferred to an electric furnace for completion of the fusion.
F, heat in an electric furnace at the temperature stated.
LiI, add a crystal of lithium iodide or iodate to the casting dish immediately before casting the melt, to aid release of the bead.

Material	Flux	Flux weight	Sample weight	Fusion	Casting	Melt weighed
Routine samples	4/1	7.5	1.5	GB+F 1200°C	From 1200°C F	N
Albite	R	R	R	R	R	N
Alumina, high purity	T	R	R	R	R	N
Alumina spinel	R	R	R	R	R	N
Aluminosilicates	R	R	R	R	R	N
Aluminous materials	R	R	R	R	R	N
Amosite	R	R	R	R	R	N
Anatase	T	9.6	0.8	R	R	Y
Andalusite	R	R	R	R	R	N
Anorthite	R	R	R	R	R	N
Asbestos[a]	R	R	R	R	R	N
Attapulgite	R	R	R	R	R	N
Baddeleyite	T	9.6	0.8	RE	R	Y
Ball clay	R	R	R	R	R	N
Barium carbonate[b]	T	9.6	1.03 CL	R	R	Y
Barytes	T	9.6	0.8	1050	1050	Y
Bauxite	R	R	R	R	R	N
Bone ash	R	R	R	R	R	Y
Bone china body	R	R	R	R	R	N
Boron carbide	T/L	4.41/2.01	1.5	Chap. 18	R	Y
Boron nitride	L/B	3.93/5.31	1.5	Chap. 18	R	Y
Brucite	T	9.0	0.9	R	R	N
Calcite	T	7.5	1.5	R	R	N
Calcium carbonate	T	7.5	1.5	R	R	N
Calcium fluoride	R	R	R	R 1050	1050	Y
Calcium phosphates	T or R	7.5	1.5	R	R	Y
Calcium silicates	R	R	R	R	R	N
Cement	R	R	R	R	R	N
Chert	R	R	R	R	R	N
China clay	R	R	R	R	R	N
Chromite	M/T	4/5	0.4	RE	R+LiI	Y
Chrome-magnesite	M/T	4/5	0.4	RE	R+LiI	Y
Chrome ore	M/T	4/5	0.4	RE	R+LiI	Y
Chrysotile	R	R	R	R	R	N
Clay	R	R	R	R	R	N
Cobalt aluminate	See Note A					
Cobalt oxide	See Note A					
Colours	See Chapter 19					
Colemanite	T or R	R	1.92	R	R	Y
Corundum	R	R	R	R	R	N
Cristobalite	R	R	R	R	R	N
Crocidolite	R	R	R	R	R	N
Diatomite	R	R	R	R	R	N
Dicalcium phosphate	R	R	R	R	R	N
Dolomite	T	7.5	1.5	R	R	N
Feldspar	R	R	R	R	R	N

APPENDIX II SPECIFIC FUSION TECHNIQUES

Table *(continued)*

Material	Flux	Flux weight	Sample weight	Fusion	Casting	Melt weighed
Routine samples	4/1	7.5	1.5	GB + F 1200°C	From 1200°C F	N
Ferro-alloys	*See Note B*					
Fireclay	R	R	R	R	R	N
Flint	R	R	R	R	R	N
Fluorspar	R	R	R	R 1050	1050	Y
Gibbsite	T	7.5	1.5	R	R	N
Glasses	*See Chapter 17*					
Glazes	*See Chapter 17*					
Graphitic SiC	*See Chapter 18*					
Gypsum	R	R	R	R 1050	R 1050	Y
Haematite	T	9.6	0.8	R 1050	1050	Y
Illite	R	R	R	R	R	N
Iron oxides	T	9.6	0.8	R	R	Y
Iron silicates	R	R	R	R	R	N
Kaolinite	R	R	R	R	R	N
Kyanite	R	R	R	R	R	N
Lead bisilicate	R	R	R	R 1050	R 1050	N
Lead borosilicates	R	R	R	R 1050	R 1050	N
Lepidolite	R	R	R	R	R	N
Limestone, lime	R	R	R	R	R	N
Magnesia	T	9.0	0.9CL	R	R	N
Magnesite, burnt	T	9.0	0.9	R	R	N
Magnesite, raw	T	9.0	0.9CL	R	R	N
Magnesium carbonate	T	9.0	0.9CL	R	R	N
Magnesium spinel	T	9.0	0.9	R	R	N
Magnesite-chrome $<20\%Cr_2O_3$	T	9.0	0.9	R	R	Y
Magnesite-chrome $>20\%Cr_2O_3$	M/T	4/5	0.4	RE	R + LiI	Y
Magnetite	T	9.6	0.8	R 1050	1050	Y
Manganese oxides	T	9.6	0.8	R	R	Y
Marble	R	R	R CL	R	R	N
Marl	R	R	R	R	R	N
Mica	R	R	R	R	R	N
Molochite	R	R	R	R	R	N
Molybdenum oxide	R	R	R	R 1050	1050	Y
Montmorillonite	R	R	R	R	R	N
Mullite	R	R	R	R	R	N
Muscovite	R	R	R	R	R	N
Nepheline syenite	R	R	R	R	R	N
Nickel oxide	T	9.0	0.5	R 1050	1050	N
Olivine	R	R	R	R	R	N
Orthoclase	R	R	R	R	R	N
Periclase	T	9.0	0.9CL	R	R	N
Petalite	R	R	R	R 1050	1050	Y
Plaster	R	R	R	R 1050	R 1050	Y

continued overleaf

Table (continued)

Material	Flux	Flux weight	Sample weight	Fusion	Casting	Melt weighed
Routine samples	4/1	7.5	1.5	GB+F 1200°C	From 1200°C F	N
Quartz	R	R	R	R	R	N
Quartzite	R	R	R	R	R	N
Rutile	T	9.6	0.8	R	R	Y
Sand	R	R	R	R	R	N
Serpentine	R	R	R	R	R	N
Silica	R	R	R	R	R	N
Silicon	L/T	2.228/6.6	0.70	See Chapter 18		
Silicon carbide	See Chapter 18					
Silicon nitride	See Chapter 18					
Sillimanite	R	R	R	R	R	N
Slags[b]	R	R	R	R	R	Y
Soapstone	R	R	R	R	R	N
Spinel	T	9.0	0.9	R	R	N
Stannic oxide	See Chapter 17					
Steatite	R	R	R	R	R	N
Talc	R	R	R	R	R	N
Titania	T	9.6	0.8	R	R	Y
Vermiculite	R	R	R	R	R	N
Vitreous enamels	See Chapter 19					
Welding fluxes	See Chapter 19					
Whiting	R	R	R	R	R	N
Witherite[b]	T	9.6	1.03CL	R 1050	1050	Y
Wollastonite	R	R	R	R	R	N
Zeolites	R	R	R	R	R	Y
Zinc oxide	T	9.6	0.8	R 1050	1050	Y
Zircon	R	R	R	R	R	Y
Zirconia	T	9.6	0.8	RE	R	Y

[a]See Note 1, Appendix I.
[b]This type of sample may contain one or more of the following: PbO, ZnO, As_2O_3, Co_3O_4, SrO, SO_3, F etc. A qualitative examination is normally necessary before starting such an analysis. Reference should be made to any text concerning specific types of material, and elsewhere, the procedure being amended as may be appropriate.

Note A

Cobalt at high concentrations is extremely difficult to fuse without some migration and alloying with the metal of the dish. It is usual to make two dilutions, one in silica and one in boric oxide or alumina: 0.3 g sample + 1.2 g SiO_2 or B_2O_3 or Al_2O_3 ('dried' basis) in 7.5 g of 4/1 flux. Fuse as glaze.

It has been found possible, with great care, to fuse cobalt aluminates or colours containing up to 45% of Co_3O_4 using 1.5 g of sample in 7.5 g of 4/1 flux at 1000°C with no apparent loss of cobalt. Similarly, cobalt oxide has been fused (and calibrated) using 0.8 g of sample in 9.6 g of lithium tetraborate at 1000°C.

Note B

Fusion conditions will depend on the nature of the alloying element. See Chapter 18.

Appendix III
Problem Elements or Oxides

Problems that may or will be encountered with a number of elements are described in detail in appropriate parts of the main text. For reference purposes, an outline of these is given below with brief comments as to how best the problems may be tackled. Fusion is regarded as being carried out at 1200°C unless otherwise specified. If more than one of the problem elements is present, it is necessary to combine the various precautions in the best manner possible.

Where the abbreviation for a releasing agent, LiI, is used, an addition of lithium iodide or iodate should be made to the casting mould immediately prior to casting.

Oxide or element	Problem	Procedure
Halogens, F, Cl, Br, I	Wholly or partly volatile	Drive off during fusion and weigh melt or fuse at 1050°C, or try to retain and weigh
Sulphur trioxide	Partially volatile	Drive off during fusion or fuse at 1050°C to retain all. Weigh melt
Chromium sesquioxide	Causes melt to stick to alloy of mould	Add LiI
Cobalt oxide	Will alloy with dish	Fuse at 1050°C, do not use gas burner
Nickel oxide	Will alloy with dish	Fuse at 1050°C, do not use gas burner
Copper oxide	Will alloy with dish, causes melt to stick	Fuse at 1050°C, use LiI
Zinc oxide	Alloys with and attacks dish	Fuse at 1050°C, do not use gas burner
Arsenic trioxide	Volatile	Fuse at 1050°C. Weigh melt
Selenium	Volatile	Will tend to volatilize whatever the fusion temperature. Weigh melt
Strontia	Carbonate stable at 1025°C	Carbonates decompose on fusion. Weigh melt
Barium oxide	Carbonate stable at 1025°C	Carbonates decompose on fusion. Weigh melt
Molybdic oxide	Volatile, will alloy with dish	Fuse at 1050°C. Weigh melt

continued

Table *(continued)*

Oxide or element	Problem	Procedure
Noble metals: Pd, Tc, Ru, Rh, Re, Os, Ir, Pt, Au	Do not convert to oxides, but alloy with fusion dish	If fused, metals will alloy with dish or float in melt and cannot be determined
Silver oxide	Alloys with dish, causes melt to stick	Add LiI, silver itself cannot be determined. AgI may precipitate
Cadmium oxide	Volatile	Fuse at 1050°C. Weigh melt
Stannic oxide	Virtually infusible (<12%)	Fuse at 1050°C, dilute above limit of fusibility
Mercuric oxide	Volatile, poisonous. Damages dish	Drive off by fusing in fume cupboard. Weigh melt
Lead oxide	If reduced, alloys with dish, with damage	Fuse at 1050°C, do not use gas burner
Rare earths	May change oxidation state on fusion	Weigh melt
Uranium oxide	Poisonous and radioactive	Can be fused as normal. *Needs handling with CARE*

Appendix IV
Certified Reference Materials

Although there are many suppliers of Certified Reference Materials, figures are given in this Appendix only for those supplied by the Bureau of Analysed Samples, Middlesbrough, England and the National Institute of Standards and Technology, Gaithersburg, Maryland, USA. In the Tables below, samples prefixed BCS are produced by the former and those prefixed SRM by the latter. Some samples are coded ECRM; these are European Community Reference Materials, and can be obtained from several suppliers including the Bureau of Analysed Samples. The figures quoted with permission are taken from the respective catalogues and are nominal values. Reference should be made to the actual certificates for the certified values.

The figures quoted in italics are best estimates but not certified. LOI in tables means loss on ignition; nd means not determined.

A.IV.1 SILICA/ALUMINA MATERIALS

Figures are given in % content. Totals are not included in the certificate, but are included here, where appropriate, as they provide an additional indication of the quality of the analyses.

Table A.IV.1.1 High-silica materials

Sample type	Silica brick	High purity silica	Silica brick	Glass sand	Glass sand	Silica refract.	Silica refract.
Sample number	BCS-CRM 267	BCS-CRM 313/1	BCS-CRM 314	SRM 81a	SRM 1413	SRM 198	SRM 199
SiO_2	95.9	99.7	96.2	nd	82.77	nd	nd
TiO_2	0.17	0.02	0.19	0.12	0.11	0.02	0.06
Al_2O_3	0.85	0.05	0.77	0.66	9.90	0.16	0.48
Fe_2O_3	0.79	0.01	0.53	0.082	0.24	0.66	0.74
CaO	1.75	0.01	1.81	nd	0.74	2.71	2.41
MgO	0.06	0.01	0.05	nd	0.06	0.07	0.13
Na_2O	0.06	0.005	0.05	nd	1.75	0.012	0.015
K_2O	0.14	<0.005	0.09	nd	3.94	0.017	0.094
Li_2O	nd	<0.0005	<0.02	nd	nd	nd	nd
BaO	nd	nd	nd	nd	0.12	nd	nd
MnO	0.15	<0.001	<0.01	nd	nd	0.008	0.007
P_2O_5	nd	nd	nd	nd	nd	0.022	0.015
Cr_2O_3	nd	<0.001	nd	46 µg/g	nd	nd	nd
ZrO_2	nd	nd	nd	0.034	nd	<0.01	0.01
LOI	nd	0.1	0.09	nd	nd	0.21	0.17
Total	99.87	99.905	99.78	—	—	—	—

Table A.IV.1.2 Feldspars

Sample type	Soda feldspar	Potash feldspar	Potash feldspar	Soda feldspar
Sample number	BCS-CRM 375	BCS-CRM 376	SRM 70a	SRM 99a
SiO_2	67.1	67.1	67.1	65.2
TiO_2	0.38	<0.02	0.01	0.007
Al_2O_3	19.8	17.7	17.9	20.5
Fe_2O_3	0.12	0.10	0.07_5	0.06_5
CaO	0.89	0.54	0.11	2.14
MgO	0.05	0.03	nd	0.02
Na_2O	10.4	2.83	2.5_5	6.2
K_2O	0.79	11.2	11.8	5.2
P_2O_5	nd	nd	nd	0.02
LOI	0.39	0.35	0.40	0.26
Total	99.92	99.85	99.945	99.612

Table A.IV.1.3 Aluminosilicate and aluminous materials (for BCS bauxite see also aluminium ores, Table A.VI.6.2)

Sample type	Bauxite Arkansas	Bauxite Surinam	Bauxite Dominican	Bauxite Jamaican	Alumina reduction grade	Obsidian rock	Basalt rock
Sample number	SRM 69b	SRM 696	SRM 697	SRM 698	SRM 699	SRM 278	SRM 688
SiO_2	13.43	3.79	6.81	0.69	0.014	73.05	48.4
TiO_2	1.90	2.64	2.52	2.38	nd	0.245	1.17
Al_2O_3	48.8	54.5	45.8	48.2	nd	14.15	17.36
Fe_2O_3	7.14	8.70	20.0	19.6	0.013	2.04	10.35
CaO	0.13	0.018	0.71	0.62	0.036	0.983	nd
MgO	0.085	0.012	0.18	0.058	0.0006	nd	nd
Na_2O	0.025	0.007	0.036	0.015	0.59	4.84	2.15
K_2O	0.068	0.009	0.062	0.010	nd	4.16	0.187
Li_2O	nd	nd	nd	nd	0.002	nd	nd
MnO	0.110	0.004	0.41	0.38	0.0005	0.052	0.167
Cr_2O_3	0.011	0.047	0.100	0.080	0.0002	nd	nd
P_2O_5	0.118	0.050	0.97	0.37	0.0002	nd	0.134
ZrO_2	0.29	0.14	0.065	0.061	nd	nd	nd
SO_3[a]	0.63	0.21	10.13	0.22	nd	nd	nd
V_2O_5	0.028	0.072	0.063	0.064	0.0005	nd	nd
ZnO	0.0035	0.0014	0.037	0.029	0.013	nd	nd
Ga_2O_3	nd	nd	nd	nd	0.010	nd	nd
BaO	0.008	0.004	0.015	0.008	nd	nd	nd
Co	0.0001	0.00009	0.0013	0.0045	nd	nd	nd
LOI	27.2	29.9	22.1	27.3	nd	nd	nd
Total	99.34	99.89	99.86	99.86	—	99.52	—

[a]SO_3 content is not added into the total.

Table A.IV.1.4 Aluminosilicates and aluminous materials (*continued*)

Sample type	Ball clay	Fire-brick	Sillimanite	Calcined bauxite	Clay plastic	Burnt refractories Al_2O_3 content approx.		
						40%	60%	70%
Sample number	BCS-CRM 348	ECRM 776-1	BCS-CRM 309	BCS-CRM 394	SRM 98a	SRM 76a	SRM 77a	SRM 78a
SiO_2	51.13	62.76	34.1	4.98	48.94	54.9	35.0	19.4
TiO_2	1.08	1.52	1.92	3.11	1.61	2.0_3	2.6_6	3.2_2
Al_2O_3	31.59	29.28	61.1	88.8	33.19	38.7	60.2	71.7
Fe_2O_3	1.04	1.43	1.51	1.90	1.34	1.6_0	1.0_0	1.2
CaO	0.173	0.31	0.22	0.08	0.31	0.22	0.05	0.11
MgO	0.305	0.48	0.17	0.12	0.42	0.52	0.38	0.70
Na_2O	0.344	0.49	0.34	0.02	0.082	0.07	0.037	0.078
K_2O	2.23	2.92	0.46	0.02	1.04	1.33	0.09_0	1.22
Li_2O	nd	0.02	*0.01*	<*0.01*	0.070	0.042	0.2_5	0.12
MnO	nd	nd	*0.03*	nd	nd	nd	nd	nd
Cr_2O_3	0.016	0.02	nd	*0.08*	0.03	nd	nd	nd
BaO	*0.04*	0.122	*0.006*	nd	0.03	nd[a]	nd[a]	nd[a]
P_2O_5	0.071	0.06	nd	0.22	0.11	0.12_0	0.092	1.3
ZrO_2	*0.03*	0.04	nd	*0.15*	0.042	0.15	0.21	0.31
LOI	11.75	*0.3*	0.1	*0.40*	12.44	*0.34*	*0.22*	*0.42*
Total	99.80	99.75	100.00	99.88	99.65	100.06	99.86	100.03

[a]Figures for SrO are: 76a 0.037%, 77a 0.009%, 78a 0.25%.

A.IV.3 RAW AND CALCINED CARBONATES

Table A.IV.2.1

Sample type	High purity magnesite	Magnesite	Dolomite	Limestone	Limestone argillaceous
Sample number	BCS-CRM 389 ECRM 751-1	BCS-CRM 319	BCS-CRM 368	BCS CRM 393	SRM 1c
SiO_2	0.89	1.55	0.92	0.70	6.84
TiO_2	0.015	0.03	<0.01	0.009	0.07
Al_2O_3	0.23	0.97	0.17	0.12	1.3
Fe_2O_3	0.29	4.63	0.23	0.045	0.55
CaO	1.66	2.28	30.8	55.4	50.3
MgO	96.7	90.6	20.9	0.15	0.42
Na_2O	0.03[a]	0.04	<0.01	<0.03	0.02
K_2O	0.01[a]	0.02	<0.01	0.02	0.28
Li_2O	<0.02	0.02	<0.01	nd	nd
MnO	0.008	0.14	0.06	0.010	0.025
Cr_2O_3	0.28	0.02	<0.01	nd	nd
BaO	nd	nd	nd	0.006	nd
B_2O_3	0.029	0.10	nd	nd	nd
SrO	nd	nd	nd	0.019	0.030
LOI	nd	nd	46.7	43.4	39.9
Total	100.14	100.4	99.78	99.88	99.74

[a]Signifies values which are currently undergoing re-certification.

A.IV.3 CHROME-BEARING REFRACTORIES

Table A.IV.3.1

Sample type	Magnesite chrome bricks			Chrome refractory
			Low silica	
Sample number	BCS-CRM 369	BCS-CRM 370	BCS-CRM 396	SRM 103a
SiO_2	2.59	3.01	1.37	4.63
TiO_2	0.14	0.13	0.26	0.22
Al_2O_3	14.7	12.3	5.73	29.96
Fe_2O_3	10.3	7.23	10.9	FeO 12.43
CaO	1.17	1.54	1.12	0.69
MgO	53.5	61.8	64.6	18.54
Na_2O	0.05[a]	0.06[a]	*0.06*	nd
K_2O	0.03[a]	0.03[a]	*0.03*	nd
Li_2O	0.03[a]	0.03[a]	*0.05*	nd
MnO	0.11	0.11	0.17	0.11
Cr_2O_3	17.2	13.4	15.6	32.06
BaO	*<0.01*	*<0.01*	nd	nd
B_2O_3	nd	nd	0.09	nd
SrO	*<0.01*	*<0.01*	nd	nd
P_2O_5	nd	nd	nd	0.01
ZrO_2	nd	nd	nd	0.01
LOI	nd	nd	*0.04*	nd
Total	99.82	99.64	100.02	98.66

[a]Signifies values which are currently undergoing re-certification.

A.IV.4 PORTLAND CEMENTS

Table A.IV.4.1

Sample type	Sulphate resisting	White	Ordinary
Sample number	BCS-CRM 353	BCS-CRM 354	BCS-CRM 372/1
SiO_2	20.5	21.8	20.3
TiO_2	0.16	*0.04*	0.29
Al_2O_3	3.77	4.85	5.3
Fe_2O_3	4.82	0.30	3.42
CaO	64.8	70.0	65.3
MgO	2.42	0.42	1.31
Na_2O	0.10	0.10	0.10
K_2O	0.49	0.11	0.76
Cr_2O_3	*0.02*	0.004	*0.012*
Mn_2O_3	0.23	0.058	0.074
SrO	0.23	0.11	*0.049*
SO_3	2.25	2.25	2.95

Additional figures on 372/1 $P_2O_5 = 0.073\%$ and $Cl = 0.008\%$.

APPENDIX IV CERTIFIED REFERENCE MATERIALS

Table A.IV.4.2 All the cements below are SRM samples

Sample type	Red 633	Gold 634	Blue 635	Yellow 636	Pink 637	Green 638	Clear 639	Black 1880	White 1881
SiO_2	21.8_8	20.7_3	18.4_1	23.2_2	23.0_7	21.4_8	21.6_1	19.82	22.25
TiO_2	0.24	0.29	0.32	0.18	0.21	0.25	0.32	0.23	0.23
Al_2O_3	3.7_8	5.2_1	6.2_9	3.0_2	3.2_8	4.4_5	4.2_8	5.02	4.19
Fe_2O_3	4.20	2.84	2.61	1.61	1.80	3.55	2.40	2.91	4.68
CaO	64.5_0	62.5_8	59.8_3	63.5_4	66.0_4	62.0_9	65.7_6	63.13	58.67
MgO	1.0_4	3.3_0	1.2_3	3.9_5	0.6_7	3.8_3	1.2_6	2.69	2.62
Na_2O	0.64	0.15	0.07	0.11	0.15	0.13	0.65	0.28	0.04
K_2O	0.17	0.42	0.45	0.59	0.25	0.59	0.06	0.91	1.17
Mn_2O_3	0.04	0.28	0.09	0.12	0.06	0.05	0.08	0.08	0.26
P_2O_5	0.24	0.10	0.17	0.08	0.24	0.06	0.08	0.29	0.09
SrO	0.31	0.12	0.21	0.04	0.09	0.07	0.15	0.06	0.11
Cr_2O_3	0.01	0.08	0.01	0.01	0.01	0.01	0.01	nd	nd
ZnO	0.01	0.02	0.01	0.03	0.01	0.10	0.01	0.01	0.01
SO_3	2.2_0	2.2_1	7.0_7	2.3_1	2.3_8	2.3_4	2.4_8	3.37	3.65
F	0.08	0.08	0.04	0.06	0.04	0.04	0.02	0.10	0.09
LOI	0.7_5	1.6_2	3.24	1.1_6	1.6_9	0.9_5	1.0_0	1.38	2.01
Total	100.06	100.00	100.03	100.00	99.97	99.97	100.16	100.30	100.07

A.IV.5 ZIRCON AND RUTILE

Table A.IV.5.1 Zircon

Sample number	BCS-CRM 388
SiO_2	32.7
TiO_2	0.25
Al_2O_3	0.33
Fe_2O_3	0.06
CaO	0.04
MgO	<0.05
Na_2O	<0.02
K_2O	<0.02
P_2O_5	0.12
ZrO_2[a]	66.2
LOI	0.2
Total	99.90

Table A.IV.5.2 Rutile

Sample number	SRM 670
SiO_2	0.51
TiO_2	96.16
Fe_2O_3	0.86
Cr_2O_3	0.23
V_2O_5	0.66
ZrO_2[a]	0.84
Total	99.26

[a]This is the sum of $ZrO_2 + HfO_2$, as some methods used cannot distinguish between these oxides, and is so reported on the certificate. In the case of BCS-CRM 388 a separate hafnia figure by XRF is given but not certified.

A.IV.6 ORES

Table A.IV.6.1 Iron ores

Sample type	Lincolnshire	Iron ore[a]	Iron ore[a]	Sibley	Canada	Labrador	Nimba
Sample number	BCS-CRM 301/1 ECRM 651-1	BCS-CRM 302/1 ECRM 681-1	BCS-CRM 175/2 ECRM 682-1	SRM 27f	SRM 690	SRM 692	SRM 693
SiO_2	7.40	17.81	0.48	4.17	3.71	10.14	3.87
TiO_2	0.30	0.48	0.053	0.019	0.22	0.045	0.035
Al_2O_3	4.26	10.62	0.384	0.82	0.18	1.41	1.02
Fe	23.85	33.21	68.74	65.97	66.85	59.58	65.11
CaO	22.6	5.03	0.025	0.039	0.20	0.023	0.016
MgO	1.73	1.48	0.029	0.019	0.18	0.035	0.013
Na_2O	0.07	0.092	*0.0046*	0.012	0.003	0.008	0.0028
K_2O	0.32	0.59	0.006	0.008	0.003	0.039	0.0028
MnO	1.25	0.28	0.274	0.011	0.23	0.46	0.091
P_2O_5	0.80[a]	2.01	0.082	0.094[a]	0.025[a]	0.089[a]	0.128[a]
S as SO_3	1.00	0.257	0.010	0.013[a]	0.008[a]	0.13[a]	0.0125[a]
Cr_2O_3	nd	0.060	nd	nd	nd	nd	nd
F	nd	0.19	*0.002*	nd	nd	nd	nd
V_2O_5	nd	0.137	nd	nd	nd	nd	nd
NiO	nd	0.020	nd	nd	nd	nd	nd
LOI	25.8			nd	nd	nd	nd

[a]Figures calculated as oxides from the quoted elemental data.
BCS 175/2 also quotes: ZnO 0.002%, PbO 0.001%, CdO 0.0003% and As_2O_5 0.015%.

Table A.IV.6.2 Manganese, chromium and aluminium ores

Sample type	Manganese ore	Manganese ore	Grecian chrome ore	Bauxite[a]
Sample number	BCS-CRM 176/2	SRM 25d	BCS-CRM 308	BCS-CRM 395
SiO_2	2.53	2.52	4.25	1.24
TiO_2	0.30	0.11	0.16	1.93
Al_2O_3	5.2	5.32	19.4	52.4
Fe_2O_3	9.81	3.92	17.0	16.3
CaO	0.09	0.052	0.34	0.05
MgO	0.04	nd	16.4	0.02
Na_2O	0.11	nd	0.04	0.03
K_2O	1.30	0.93	0.01	0.01
MnO	61.3	51.78(Mn)	0.14	<0.02
Cr_2O_3	nd	nd	41.5	0.07
BaO	0.19	0.21	nd	nd
SO_3	0.045	0.13	nd	nd
P_2O_5	0.199	0.25	nd	nd
As_2O_3	0.22	nd	nd	nd
PbO	0.01	nd	nd	0.003
LOI	nd	(H_2O)0.96	0.9	27.8
Total	—	—	100.14	99.853

[a]For other bauxites see Table A.IV.1.3.

Table A.IV.6.3 Fluorspar

Sample number	BCS-CRM 392
SiO_2	0.67
CaO[a]	0.52
BaO	0.37
PbO	0.18
CO_2	0.48
SO_3	0.30
CaF_2	97.2

[a]The CaO quoted here is the 'free' CaO taken as being present as the carbonate as distinct from that taken as combined in CaF_2. It is usually determined from the acetic acid-soluble material.

A.IV.7 BASIC SLAGS

Table A.IV.7.1

Sample number	BCS-CRM 381	BCS-CRM 382/1 ECRM 851-1	ECRM 879-1
SiO_2	8.78	13.03	8.82
TiO_2	0.35	0.42	0.535
Al_2O_3	0.67	3.79	0.803
Fe	13.3	19.9	18.97
FeO	3.69	nd	nd
CaO	49.0	40.1	43.70
MgO	1.03	3.73	2.19
MnO	3.16	7.96	4.45
Cr_2O_3	0.33	0.80	0.447
P_2O_5	15.7	3.06	8.46
V_2O_5	0.94	0.24	0.738
F	nd	0.10	0.368
S	0.19	0.37	0.102

A.VI.8 PHOSPHATE ROCKS

Table A.IV.8.1

Sample type	Florida SRM 120b	Western SRM 694
SiO_2	4.68	11.2
TiO_2	0.15	*0.11*
Al_2O_3	1.06	1.8
Fe_2O_3	1.10	0.79
CaO	49.40	43.6
MgO	0.28	0.33
Na_2O	0.35	0.86
K_2O	0.12	0.51
P_2O_5	34.57	30.2
F	3.84	3.2
CdO	0.002	0.015
Cr_2O_3	nd	*0.10*
U ($\mu g/g$)	128.4	141.4

Figures also given on 694 are: V_2O_5 0.31%, ZnO *0.19%*.

A.IV.9 FERRO-ALLOYS

Table A.IV.9.1

Sample type	Ferrotungsten	Ferrotungsten	Ultra-low C ferrochromium	Ferroniobium (40% Nb)	Ferrovanadium
Sample number	ECRM 590-1	BCS-CRM 242/2 ECRM 555-1	BCS-CRM 366 ECRM 554-1	ECRM 576-1	BCS-CRM 205/3 ECRM 577-1
C	0.025	0.025	0.007	0.201	0.089
Si	1.05	1.75	0.53	1.79	1.79
Mn	0.136	nd	0.09	nd	0.158
P	nd	0.02	0.018	nd	0.035
S	nd	0.018	<0.005	nd	0.034
Cr	nd	nd	74.6	nd	nd
Ni	nd	nd	0.28	nd	0.053
Al	0.36	0.14	nd	2.53	0.414
Co	nd	nd	0.043	nd	nd
Cu	0.0484	nd	nd	nd	0.054
Nb	nd	nd	nd	43.9	nd
Sn	0.045	0.034	nd	0.195	nd
Ti	nd	nd	nd	1.32	nd
V	nd	nd	0.10	nd	50.16
W	79.55	79.9	nd	nd	nd
Fe	nd	15.2	nd	nd	nd
Mo	0.101	nd	nd	nd	nd

Table A.IV.9.2

Sample type	Ferro-molybdenum	Ferro-niobium (60% Nb)	Low C ferro-chromium	Ferro-manganese	Ferro-titanium	Ferro-chromium (charge)	Ferro-boron
Sample number	BCS-CRM 231/4 ECRM 578-1	BCS-CRM 362 ECRM 579-1	BCS-CRM 203/5 ECRM 580-1	BCS-CRM 280/2 ECRM 583-1	BCS-CRM 243/4 ECRM 584-1	ECRM 585-1	BCS-CRM 363 ECRM 587-1
C	0.016	0.037	0.019	0.333	0.044	6.87	0.738
Si	0.208	1.03	0.306	0.396	1.80	2.76	0.129
Mn	nd	nd	nd	86.42	1.13	0.86	0.272
P	0.024	0.064	0.011	0.146	0.032	0.018	0.020
S	0.065	0.021	nd	0.007	0.030	0.039	0.001
Cr	nd	nd	72.18	nd	nd	57.6	0.108
Mo	72.23	nd	nd	nd	nd	nd	0.005
Ni	nd	nd	nd	nd	nd	0.197	nd
Al	nd	1.86	nd	nd	7.19	nd	0.047
B	nd	nd	nd	nd	nd	nd	18.7
Co	nd	0.005	0.047	nd	nd	0.044	0.010
Cu	0.136	nd	nd	nd	nd	nd	nd
Nb	nd	62.87	nd	nd	nd	nd	nd
Sn	nd	0.344	nd	nd	nd	nd	nd
Ti	nd	0.567	nd	nd	37.17	0.36	0.04
V	nd	nd	0.083	nd	nd	0.33	0.004
Ta	nd	3.85	nd	nd	nd	nd	nd
Fe	nd	nd	nd	12.3	nd	31	nd

APPENDIX IV CERTIFIED REFERENCE MATERIALS

A.IV.10 GLASSES

Table A.IV.10.1 Soda-lime glasses

Sample type	Flat SRM 620	Container SRM 621	Float SRM 1830	Sheet SRM 1831
SiO_2	72.8	71.13	73.07	73.08
TiO_2	0.018	0.014	0.011	0.019
Al_2O_3	1.80	2.76	0.12	1.21
Fe_2O_3	0.043	0.040	0.121	0.087
CaO	7.11	10.71	8.56	8.20
MgO	3.69	0.27	3.90	3.51
Na_2O	14.39	12.74	13.75	13.32
K_2O	0.41	2.01	0.04	0.33
As_2O_3	0.056	0.030	nd	nd
SO_3	0.28	0.13	0.26	0.25
Total	*100.60*	*99.83*	*99.83*	*100.01*

Other figures on 621 are: ZrO_2 0.007%, BaO 0.12%

Table A.IV.10.2 Glasses, various

Sample type Sample number	Lead–barium SRM 89	Opal SRM 91	Low-boron SRM 92	High-boron SRM 93a	Borosilicate (soft) SRM 1411	Multi-component SRM 1412
SiO_2	65.35	67.50	75.0	80.8	58.04	42.38
TiO_2	0.01	0.019	nd	0.014	0.02	nd
Al_2O_3	0.18	6.01	nd	2.28	5.68	7.52
Fe_2O_3	0.049	0.079	nd	0.028	0.050	*0.031*
CaO	0.21	10.49	8.3	0.01	2.18	4.53
MgO	0.03	*0.008*	0.1	0.005	0.33	*4.69*
Na_2O	5.70	8.47	*13.1*	3.98	10.14	4.69
K_2O	8.40	3.24	*0.6*	0.014	2.97	4.14
Li_2O	nd	nd	nd	nd	nd	*4.57*
SO_3	0.03	nd	nd	nd	nd	nd
PbO	17.50	0.10	nd	nd	nd	4.40
ZnO	nd	0.08	*0.2*	ns	3.85	4.48
CdO	nd	nd	nd	nd	nd	4.38
MnO	0.088	*0.008*	nd	nd	nd	nd
ZrO_2	0.005	0.009	nd	0.042	nd	nd
BaO	1.40	nd	nd	nd	5.00	4.67
B_2O_3	nd	nd	0.70	12.56	10.94	4.53
P_2O_5	0.23	0.023	nd	nd	nd	nd
As_2O_5	0.36	0.10	nd	nd	nd	nd
As_2O_3	0.03	0.09	nd	nd	nd	nd
Cl	0.05	0.015	nd	0.06	nd	nd
SrO	nd	nd	nd	nd	0.09	4.55
F	nd	5.73	nd	nd	nd	nd
LOI	0.32	nd	*0.42*	nd	nd	nd
Total	*99.942*	*101.971*	*98.42*	*99.793*	*99.29*	*99.561*

A.IV.11 FURNACE DUST

Table A.IV.11.1

Sample number	ECRM 877-1
Si	1.08
Ti	0.032
Al	0.044
Fe	62.07
Ca	3.23
Mg	0.28
Na	0.23
K	0.058
Mn	1.37
Cr	0.017
P	0.18
S	0.18
F	0.78
V	0.029
Ni	0.010
C	0.83
Zn	1.16
Pb	1.00
Cd	*0.003*
Cu	0.025
Cl	*0.031*
As	0.014

Appendix V
Laboratory Accreditation

There are a great many laboratories offering analytical and other testing services. It has been known for a considerable time that not all of these can be relied upon to produce accurate results. National and international commerce has come to rely, to an increasing extent, on specification testing by external laboratories as a means of acceptance testing, this being an alternative to testing in the purchaser's or supplier's own facilities. It has the great advantage of reducing the likelihood of disputes arising, provided that both parties accept the integrity and accuracy of the independent laboratory. Frequently, in the past, discrepant results have resulted in goods being returned, and in discord between supplier and customer, and possibly in extreme cases litigation. In many such cases, common sense between the disputing parties led to an agreement to have conducted a 'referee' analysis by a reputable laboratory, and for the results obtained to be accepted as definitive. The Ceram Research's Analytical and Testing Services laboratory was commonly accepted as such an establishment where one of the parties originated in the United Kingdom.

Such arrangements, in the past, were dependent on the established reputation of the laboratory chosen, which meant that, usually, the laboratory was known personally to at least one of the parties. It was realized that this was not an ideal arrangement and that some form of official list of suitable laboratories should be available. On a national scale, this necessitated some form of government agency being set up. This, in the United Kingdom, was NATLAS (National Accreditation of Testing Laboratories Scheme) now re-named NAMAS (National Measurement Accreditation Scheme). Under such a scheme no laboratory can be included which does not measure up to approved standards. Such standards and the arrangements for assessing the performance of laboratories are incorporated in British Standards (BS 5781, Part 1:1979 and Part 2:1981, and BS 6460) and will eventually be included in Euro Standard EN 45001.

Thus, any laboratory wishing to become accredited has to meet approved and stringent standards, not only in respect of the actual quality of the work and the system employed to control its quality but also with reference to administrative detail. This entails a thorough inspection of the establishment by a team from the responsible authority consisting of individual experts covering various technical fields, laboratory administration and the system of application of the scheme itself before an accreditation certificate is issued. This is followed by further checks annually on random, but specific, parts of the system, both technical and administrative. The accrediting authority needs to satisfy itself from inception onwards that analytical and other test results are up to standard. In addition, the administration must be adequate to ensure correct derivation of results from the received samples and, finally, personnel must be adequate to the task and must be kept fully informed of any changes in methodology or system.

Needless to say, wherever possible the accrediting authority prefer methods of test to be officially approved national or international standard methods, and require the equipment etc. used to be capable of reference to appropriate standards. Where the former proviso is not the case, they need to be satisfied that the methods to be used can be referred to published work of acceptable calibre or, failing this, that they are fully documented and proved. Any documentation must be readily available to all with a right of access, and must be kept up to date. If used by an accredited laboratory, the methods as described in this book are in a format that should be acceptable to NAMAS or similar organizations as written down versions of methods, and could avoid the need for separate documentation.

Great stress is laid on derivation, accountability and referability to *prime* standards. Similar restrictions apply to the measurement of time, volume, pressure, flow-rates etc. Calibration of all equipment is another essential, repeat calibrations having to fall within acceptable intervals—once again by reference to approved systems. Weights, for example, in the case of NAMAS need to be referred to a set calibrated originally by the UK National Physical Laboratory, and to be checked periodically either against such a set or, where appropriate, by the local Weights and Measures Authority or other acceptable organization. It is necessary to keep records of these arrangements, on paper or computer, to show adherence to the agreed schemes. Failure to comply with any of these requirements can result in the suspension of accreditation for that activity, either temporarily or permanently.

The accrediting authority may also require check tests or analyses to be carried out whenever such a request is made; this does not appear to be invoked in most activities. The power, however, is there.

One further requirement of the accrediting authority is that the laboratory can demonstrate that it is 'independent' in the sense that customers can be assured that it cannot be subjected to pressures that may possibly restrict its ability to report honest results, honestly obtained.

It should, therefore, be possible to assume that employment of an accredited laboratory will yield definitive answers. Even so, the accrediting authority is prepared to receive any complaint about the conduct of, or the results achieved by, any laboratory in the scheme. It would then investigate the complaint, and has the power to remove the offending laboratory from the list, if found guilty.

This NAMAS scheme is the UK national system, but similar schemes exist or are being created in many other countries. As trade becomes increasingly international, considerable efforts are being made to obtain acceptance of agreements to harmonize standards and conditions by bi- and multilateral agreements. The ultimate objective is that results from any nationally accredited laboratory will be accepted internationally. This would obviate the need either to dispatch batch samples for testing in the purchaser's nationally accredited laboratory, still leaving possible arguments that sample and batch differ, or to dispatch the goods and have them tested on receipt, and possibly rejected. Several mutual acceptance agreements have already been achieved, and the EEC is in the process of establishing a common accreditation system for the whole Community. In the UK, much of this work in the international field is undertaken under the auspices of BSI and ISO.

The movement towards accreditation has received a further spur with the establishment of accredited quality assurance schemes, e.g. BS 5750. Testing of raw materials and products by an accredited laboratory is regarded as a desirable feature within such a proposed scheme. It is also clearly an argument for acceptance of the goods by prospective customers.

There is much to be said for a laboratory achieving accredited status; there is equally as much to be said for a manufacturer using an accredited laboratory for specification testing, or a purchaser insisting on using one for that purpose. Apart from the obvious merits conferred by the standards of quality demanded by the scheme, the laboratory has the advantage of being independent. It was not unknown when chemical methods

APPENDIX V LABORATORY ACCREDITATION

were being used for techniques to be selected that were known to produce results where the known bias would favour the interests of the analyst's employers. Some continental countries, in fact, tended to lump together alumina and titania as a single figure, as the higher the apparent alumina content of an aluminosilicate or aluminous material the better.

British Ceramic Research Limited was one of the first to join the NATLAS scheme (it was, in fact, No. 13 in the list), as it recognized not only the importance to itself but also the advantages that would accrue to its members and customers who might wish to use its services. It took a great deal of trouble to ensure that its full range of XRF methods was accredited, and has also done a great deal to help to produce the start of a series of standard methods of analysis of refractory materials by the fused, cast bead technique with an XRF finish. Thus, it has proved possible to accredit XRF analysis using the techniques described in this book, covering a vast range of ceramic, mineralogical and geochemical needs. This should be of great advantage to all involved in such types of analysis, as the use of approved methods of analysis must enhance the status of the results.

Index

(Page numbers in **bold** type indicate main point of discussion)

Many materials not mentioned specifically in the Index may be referred to in Appendices I and II

Accreditation, laboratory, 36, **287**
Albite, detailed procedure, 193
α-coefficient, *see* Inter-element effect
Alumina, line choice and occurrence, 108
Alumina 'Pure', detailed procedure, 194
Aluminosilicate materials,
   composition of CRMs, 275–276
   detailed procedure, **193**
Aluminous materials
   composition of CRMs, 275–276
   detailed procedure, **193**
Analytical methods, for specific materials or types, see under specific materials or types, 193
Anatase, detailed procedure, 227
Andalusite, detailed procedure, 194
Anhydrite, detailed procedure, 206
Anorthite, detailed procedure, 201
Antimony,
   as reducible element, 89
   as volatile element, 90
   trioxide, line choice and occurrence, 131
Apatite, *see* Bone ash
Arsenic,
   as reducible element, 89
   as volatile element, 90
Arsenious oxide, line choice and occurrence, 124
Asbestos, safety aspects, note, 265
Automatic fusion apparatus, *see* Fusion

Baddeleyite, detailed procedure, 227
Balances, **22**
   electronic, 23
   mechanical, 22
Ball clay,
   composition of CRM, 277
   detailed procedure, 193

Barium oxide, line choice and occurrence, 132
Basalt, composition of CRMs, 276
Bauxite,
   composition of CRMs, 275, 276, 281
   detailed procedure, 194
   flux and calibration, 171
BCeramRA, Working Groups, 3
Bead,
   preparation, procedure, **148**
   rippling, 152
   standard, *see* Standard beads
   sticking, due to reduction? 89
Bead weight, effect of, **79**
   adhesion to dish, 80
   age of dish, 81
   melt temperature, 79
   viscosity of the melt, 80
Bentonite, *see* Montmorillonite
Bismuth oxide, line choice and occurrence, 143
BISRA, Working Group, 3
Blank determinations, *see* Reagents
Bone ash,
   detailed procedure, 204
   flux and calibration, 171, 172
   retention of hydroxyl, etc., 57
Bone china body,
   detailed procedure, 203
   flux and calibration, 171, 172
Borate, determination of, **41**
Boric acid, loss of B_2O_3 on heating, 54
Boron carbide,
   detailed procedure, 240, 248
   for grinding, *see* Vials
Boron nitride, detailed procedure, 240, 248
Brick, building, detailed procedure, 194

Bromine,
  line choice and occurrence, 125
  presence of in kiln deposits, 62

Cadmium,
  as volatile element, 90
  oxide, line choice and occurrence, 130
Caesia, line choice and occurrence, 132
Calcium carbonates, detailed procedure, 210
Calcium fluorides, detailed procedure, 208
Calcium oxide, see Lime
Calcium phosphates, detailed procedure, 204
Calcium-rich materials, detailed procedure, **201**
Calcium silicates,
  detailed procedure, 201
  flux and calibration, 171, 172
Calcium sulphates, detailed procedure, 206
Calculation of results, **179**
Calibration, **155, 158**
  α-correction, 156
  advantages of using synthetic standards, 68
  background correction, 156
  curves, checking, 162
  curves, establishing, **158**
  inter-element correction, 156
  inter-element correction, determination of, **166**, 169
  limestone base, 172
  line interference, 156
  line interference, determination of, **165**
  line overlap, 156
  magnesia base, 172
  range, normal conditions, 190
  see also specific materials
  silica/alumina base, 171
  standard beads, see Standard beads
  systems of, **170**
  zirconia base, 173
Carbides, detailed procedure, 240, 242, 243
Carbonate materials,
  composition of CRMs, **277**
  decomposition of by heat, **58**
Carbon, determination of, **43**
Casting, **79**
  comparison of top and bottom surfaces, 79
  effect of bead weight, 79
  see also Bead weight
  moulds, see Platinum ware
  'normal' conditions, 189
  see also specific materials
  procedure, **151**
Cement, Portland,
  calibration, 171, 172
  composition of CRMs, 278–279
  detailed procedure, 201
  flux and calibration, 171, 172
Ceramic Chemists Conferences, **2**
Cerium oxide, line choice and occurrence, 135
Certified reference standards, appendix, analyses of, **273**
***SEE ALSO* SPECIFIC MATERIALS AND TYPES**
Chalk, detailed procedure, 210
Chemical changes on heating, see Loss on ignition
China clay, detailed procedure, 193
Chlorine,
  line choice and occurrence, 116
  presence of in Middle East samples, 62
Chrome-bearing materials,
  composition of CRMs, 279, 284
  detailed procedure, 219
  fusion of, **86**
Chrome-magnesite refractory,
  composition of CRM, 278
  detailed procedure, 219
Chrome ore,
  composition of CRM, 281
  detailed procedure, 219
Chromium sesquioxide, line choice and occurrence, 118
Clay,
  burning off organic matter, **58**
  detailed procedure, 193
Clay, flint, detailed procedure, 194
Clay, plastic, composition of CRM, 276
Cobalt,
  as reducible element, 89
  oxide, line choice and occurrence, 120
Colemanite, loss of B_2O_3 on heating, 22, 54
Collimators, **102**
Colouring oxides, 256
Colours, glass, pottery, detailed procedure, 251, 252

INDEX

Computer, hardware, software, 35
Contamination, by grinding, 67
Concentration limited elements, 88, **90**
Copper,
   as concentration limited element, 91
   as reducible element, 89
   oxide, line choice and occurrence, 122
Cornish stone, detailed procedure, 193
Cristobalite, detailed procedure, 194
Crystals, *see* Spectrometer, choice of
   ammonium dihydrogen phosphate, germanium, indium antimonide, layered, lithium fluoride, pentaerythritol, thallium acid phthalate, **102**
Cupric oxide, *see* Copper

Decomposition temperatures, 50
Desiccants, **22**
   calcium chloride, magnesium perchlorate (Anhydrone), phosphorus pentoxide, silica gel, 22
Detectors, **103**
   argon/methane flow proportional, helium/CO_2 flow proportional, krypton sealed, scintillation, xenon sealed,
   see also Spectrometer, choice of
Determination of,
   individual oxides, etc., *see* materials
   loss on ignition, *see* Loss on ignition
Dicalcium phosphate, detailed procedure, 204
Dilution by fusion, *see* Fusion
Dolomite,
   composition of CRM, 277
   detailed procedure, 210
   flux and calibration, 172
Dolomitic limestone, detailed procedure, 210
Drying, **21**
   cabinets for standard beads, 22
   drying ovens, 21
   microwave ovens, 22
   'normal' conditions, 186
   see also specific materials
Dust, furnace, composition of CRM, 286

Earthenware, detailed procedure, 194
Element, *see also* specific elements/oxides
   line choice and occurrence, **105**

Evolution of volatiles, *see* Loss on ignition

Feldspar,
   composition of CRMs, 274
   detailed procedure, 193
Ferric oxide, ferrous oxide, *see* Iron
Ferro-alloys,
   composition of CRMs, 283–284
   detailed procedure, 240, 247
Firebrick,
   composition of CRM, 277
   detailed procedure, 194
Fireclay, detailed procedure, 193
Flint, detailed procedure, 194
Flint clay, detailed procedure, 194
Fluorine,
   determination of, **44**
   evolution of during heating, **63**
   line choice and occurrence, 113
Fluorspar,
   composition of CRM, 281
   detailed procedure, 208
   flux and calibration, 171, 172
Flux,
   appendix, for material types, 267
   batch, change of, 163
   choice of, **69**
   choice of, theory for, **70**
   mixed 4:1, advantages of, **71**
   welding, detailed procedure, 251, 252, 255
Frit, detailed procedure, 231
Furnace dust, composition of CRM, 286
Furnaces, *see* Heating, 25
Fusion,
   apparatus, automatic, platinum for, 29
   appendix, techniques for material types, **267**
   calibration using synthetic standards, 68
   concentration limited elements, 88, **90**
   decomposition of samples by, **102**
   dilution of sample, 68, **72**, 91
   dilution, effects of, **91**
   flux,
      see also Flux,
   iron, loss of, during fusion, 85, 88
   melt weight factor, 76
   non-oxide materials, **81**
   'normal' conditions, 188
   see also specific materials

Fusion (*continued*)
  problem elements, **88**
  procedure, **149**
  reasons for, 67
  reduced materials, **81**
  reducible elements, **88**
  silicon metal, dangers of, 83, 246
  synthetic standards, 68, **75**
  systems of, **170**
  tungsten, loss of during fusion, 85
  volatile elements, 88, **90**
  volatilization during, **77**
  see also specific types
Gadolinium oxide, line choice and occurrence, 136
Gallium oxide, line choice and occurrence, 123
Ganister, detailed procedure, 194
Germanium oxide, line choice and occurrence, 124
Glass, *see* Glaze
Glasses, composition of CRMs, 285
Glasses, special, detailed procedure, 252, 255
Glass sand, composition of CRMs, 274
Glaze, glass
  coloured, colourless, white, 232
  detailed procedure, 231
  flux and calibration, 172
Grinding,
  contamination by, 67
  see also Sample preparation
Gypsum,
  detailed procedure, 206
  drying problems, 22

Hafnia, line choice and occurrence, 137
Haematite, detailed procedure, 227
Halogens, determination of, **44**
Hausmanite, detailed procedure, 227
Heating, **24**
  blast burners, 25
  furnaces, electric, **25**
  gas burners, 24
  heating elements, electric, care of, 26
  microwave ovens, 22, 26
  supports for platinum ware, 25
  swirler, 4-unit, 24
  tunnel kiln for LOI, 26, 59
High-alumina materials,
  detailed procedure, **193**

flux and calibration, 171
High silica materials,
  composition of CRMs, 274
  detailed procedure, **193**
  flux and calibration, 171

Iceland spar, detailed procedure, 210
Illite, illitic clays, detailed procedure, 193
Ilmenite, detailed procedure, 227
Inter-element correction (α-correction), 156
  determination of, **166**, 169
Iodine, line choice and occurrence, 132
Iron,
  as reducible element, 88
  loss of during fusion, 85, 88
  oxides, line choice and occurrence, 109
Iron ores, composition of CRMs, 283
Iron oxides, detailed procedure, 227

Kaolin, detailed procedure, 193
Kiln deposits, detailed procedure, 252, 259
Kiln, tunnel, for LOI, 26, 59
Kyanite, detailed procedure, 194

Laboratory accreditation, **287**
Lanthanum oxide, line choice and occurrence, 135
Lead,
  as reducible element, 89
  oxide, line choice and occurrence, 141
Lead borosilicate, flux and calibration, 172
Lepidolite, detailed procedure, 193
Lime, line choice and occurrence, 110
Limestone,
  base calibration, 172
  composition of CRMs, 277
  detailed procedure, 210
Limit of detection, 190
Limonite, detailed procedure, 227
Line, element, **105**
  see also specific elements choice of
Line interference, 156
  determination of, **165**
Lithia, determination of, **44**
Loss on ignition, **47**
  appendix, methods for material types, **261**
  calculation of true figure, 181
  carbon dioxide, evolution of, 53, **58**
  chemical changes on heating, 50
  determination of, 48

INDEX
295

effects of composition, fineness, etc., 51
fluorine, evolution of, 53, **63**
halogens, evolution of, **62**
'normal' conditions, 187
  see also specific materials
problems with, 51
procedure, **146**
reasons for determination of, 49
sulphur gases, evolution of, 53, **60**
tunnel kiln for, 26, 59
volatiles, evolution of, **55**
water, evolution of, 55

Magnesia,
  base calibration, 172
  detailed procedure, 215
  flux and calibration, 172
  line choice and occurrence, 110
Magnesia based abrasives, detailed procedure, 215
Magnesite, see also Magnesia
Magnesite, composition of CRMs, 277
Magnesite-chrome refractory,
  composition of CRMs, 278
  detailed procedure, 215
Magnesium silicates, detailed procedure, 217
Magnetite, detailed procedure, 227
Manganese ores, composition of CRMs, 281
Manganese oxides, detailed procedure, 227
Manganic oxide, line choice and occurrence, 119
Marl, detailed procedure, 193
Masks,
  instrument, contamination of, 101
  sample, 101
  see also Spectrometer, choice of,
Melt weight factor, see Fusion
Methods, analytical, see **Procedures**
Mica, detailed procedure, 193
Microwave ovens,
  for drying, 22
  for loss on ignition, 26
Molochite, detailed procedure, 194
Molybdenum trioxide, line choice and occurrence, 130
Montmorillonite, bentonite
  detailed procedure, 193
  loss of water with temperature, 56

Mullite, detailed procedure, 194
Muscovite, detailed procedure, 193

NAMAS scheme, 287
NATLAS scheme, 287
Neodymium oxide, line choice and occurrence, 136
Nepheline syenite, detailed procedure, 193
Nickel oxide, line choice and occurrence, 121
Niobium pentoxide, line choice and occurrence, 129
Nitrides, detailed procedure, 240, 242, 245
Non-oxide materials, fusion of, **81**
Non-XRF elements, **41**
  determination of, see also individual oxides, 41
  light elements, 39
  reducible oxides, 40, 45
  volatile elements, 39, 45
'Normal' conditions, **186**
  see also specific materials

Obsidian rock, composition of CRMs, 275
Occurrence, see specific elements/oxides
Ores, composition of CRMs, **280–281**
Orthoclase, detailed procedure, 193
Oxides,
  detailed procedure, 227
  to be determined, 'normal' conditions, 189
  see also specific materials

Petalite, detailed procedure, 193
Phosphate rock,
  composition of CRMs, 282
  detailed procedure, 204
Phosphorus pentoxide, line choice and occurrence, 114
Pipes, clay, detailed procedure, 194
Plaster (of Paris), detailed procedure, 206
Platinum ware, **27**
  alloy (95% Pt/5% Au), 27
  care of, 29
  casting mould, 27, 30
  cleaning, 30
  combined fusion/casting dish, 29
  damage due to reduction, 31
  fusion dish, 27
  platinum tipped tongs, 30
  size and shape, 43
  supports for, 38

Porcelain, detailed procedure, 194
Portland cement,
  composition of CRMs, 278-279
  detailed procedure, 201
Potash, line choice and occurrence, 111
Potash feldspar,
  composition of CRMs, 274
  detailed procedure, 193
Praseodymium oxide, line choice and occurrence, 135
Problem elements,
  fusion of, 88
  appendix, table of, **271**
Procedure,
  bead preparation, **149**
  British standard, 146
  calculation of results, **179**
Procedure, continued,
  calculation of true LOI, 181
  casting, **151**
  cooling, **151**
  correction of errors due to WC, 180
  fusion, **149**
  identifying the sample, 177
  loss on ignition, **146**
  presentation of bead, 153, **177**, **178**
  routine, see Routine procedure
  standard, routine, **145**
Procedures, analytical, for specific materials or types, see *under* **specific materials or types, 193**
Psilomelane, detailed procedure, 227
Pulse-height settings, **104**
Pyrolusite, detailed procedure, 227

Qualitative analysis, see Semi-quantitative analysis
Quality assurance,
  see also Accreditation, 36
Quartz, quartzite, detailed procedure, 194
Quicklime, detailed procedure, 210

Rare earths, see specific oxides, **134**
Reagents, blank determinations, 36
Reduced materials,
  detailed procedure, **239**
  fusion of, **81**
  see also specific types
Reducible elements, **88**
  determination of, 45
Refractories, silica/alumina, burnt,
  composition of CRMs, 277
Reporting of results
  'normal' conditions, 190
  see also specific materials
  reporting limit, 190
Rippling, 152
Routine procedure as for silica/alumina, **185**
Rubidia, line choice and occurrence, 125
Rutile,
  composition of CRM, 282
  detailed procedure, 227

Samarium oxide, line choice and occurrence, 136
Sample changer, see Spectrometer, choice of
Sample holder, see Spectrometer, choice of
Sample identification, **177**
Sample preparation, **14**
  boron carbide, 15
  contamination, 15, 67
    see also Tungsten carbide
  contamination, cross, 15, 18
  drying of samples, 14
    see also Drying
  grinding, **15**
  jaw crushing, 14
  'normal' conditions, 186
    see also specific materials
  splitting, **19**
  Tema mill, **16**
  tungsten carbide, 15
    see also Vials
  vials, see Vials
Sampling, see Sample preparation, 13
Sand, glass, composition of CRMs, 274
Sand, silica, detailed procedure, 194
Scandium oxide, line choice and occurrence, 117
Selenium, determination of, **45**
Semiquantitative composition, detailed procedure, 233, **253**
Shale, detailed procedure, 193
Sialons,
  detailed procedure, 240, 247
  fusion of, **81**, 83
Siderite, ferrous carbonate,
  detailed procedure, 227
  loss of CO_2 on heating, 58

INDEX

Silica,
　flux and calibration, 171, 172
　line choice and occurrence, 107
Silica/alumina materials,
　composition of CRMs, **274**
　detailed procedure, **193**
　flux and calibration, 171, 172
Silica, brick,
　composition of CRMs, 274
　detailed procedure, 194
Silica, fused, detailed procedure, 194
Silica, high purity, composition of CRM, 274
Silica refractory, composition of CRMs, 274
Silicon carbide,
　composition of, 81
　detailed procedure, 240, 242, 243
　determination of LOI, 82
　fusion of, **81**, 83
Silicon metal,
　dangers of, during fusion, 83, 246
　detailed procedure, 240, 242, 246
Silicon nitride,
　detailed procedure, 240, 242, 245
　fusion of, **81**, 84
Sillimanite,
　composition of CRM, 276
　detailed procedure, 194
Slags,
　detailed procedure, 251, 252, 258
　flux and calibration, 171
Slags, basic, composition of CRMs, 282
Slaked lime, detailed procedure, 210
Soapstone, *see* Talc
Soda, line choice and occurrence, 112
Soda feldspar,
　composition of CRMs, 274
　detailed procedure, 193
Spectrometer, choice of, **31**
　crystals, 35
　detectors, 34
　energy dispersive, 31
　masks, 34
　sample changer, 34
　sample holder, 35
　sequential, 32
　simultaneous, 32
　simultaneous/sequential, 33
Standard beads, 157
　composition of, **160**

　drying cabinets for, 22
　presentation, 163
　ratio standards, 163
Standard methods, national and international, 174
Standards, **157**
　materials, for use as, 157
　materials, pre-treatment of, **157**
　ratio, 163
Stannic oxide, line choice and occurrence, 131
Steatite, *see* Talc
Stoneware, detailed procedure, 194
Strontia, line choice and occurrence, 126
Sulphates, decomposition of by heat, 61
Sulphur trioxide,
　as volatile oxide, 90
　determination of, **42**
　evolution of, 53, **60**
　line choice and occurrence, 115
Synthetic standards, 68, 75
　calibration, direct by using, 68
　fusion, allowing use of, 68
　pre-treatment of materials for, 78
　see also Standards

Talc,
　detailed procedure, 217
　flux and calibration, 171
Tantalum oxide, line choice and occurrence, 139
Tema mill, *see* Sample preparation
Thoria, line choice and occurrence, 143
Tiles, wall, floor, roofing, detailed procedure, 194
Tin oxide,
　as concentration limited element, 90
　see also Stannic oxide
Titanates,
　detailed procedure, 227
　flux and calibration, 173
Titania,
　detailed procedure, 227
　flux and calibration, 173
　line choice and occurrence, 108
Tricalcium phosphate, detailed procedure, 204
Tungsten, loss of during fusion, 85
Tungsten carbide, correction of errors due to, **180**

Tungstic oxide, line choice and occurrence, 140
Tunnel kiln, for LOI, 26, 59
Typical materials,
  'normal' conditions, 186
  see also specific materials

Unknown materials, detailed procedure, 251, 252
Uranium oxide, line choice and occurrence, 143

Vanadium pentoxide, line choice and occurrence, 118
Vials,
  agate, 16, 19
  cleaning, **18**
  cobalt bonding of WC vials, 17
  nickel bonding of WC vials, 17
  tungsten carbide, **15, 16**
  wear, 18
Vitreous enamels, detailed procedure, 251, 252, 255
Volatile elements, 88, **90**
  determination of, 45
Volatilization, see Fusion

Weighing, 36
  balances, see Balances
  desiccants, 22
  'normal' conditions, 187
  see also specific materials
Welding fluxes, detailed procedure, 251, 252, 255
Windows, flow counter, **103**
Working Groups,
  BCeramRA, 3
  BISRA (British Iron and Steel RA), 3

XRF,
  as standard method, 69
  relation to other instrumental techniques, 8
XRF methods,
  development of, historical, **2**
  effect of on reported results, **5**
  standardizing, 5
XRF spectrometer, 95
  ARL 8480, Philips 1410, Philips 1606, Telsec TXRF, 95
  energy dispersive, 31
X-ray tubes, **96**
  chromium, 96
  power settings, **98**
  rhodium, 97
  scandium, 97

Ytterbium oxide, line choice and occurrence, 137
Yttrium oxide, line choice and occurrence, 127

Zinc, as reducible element, 89
Zinc, oxide, line choice and occurrence, 122
Zircon,
  composition of CRM, 279
  detailed procedure, 223
  flux and calibration, 173
Zirconia, base calibration, 173
  line choice and occurrence, 128
Zircon-bearing materials, detailed procedure, 223
Zirconia, detailed procedure, 227
Zirconium silicate, detailed procedure, 223